AGRICULTURE ISSUES AND POLICIES

AGRICULTURAL ISSUES

POLICIES, CONSERVATION AND FARM PROGRAMS

Agriculture Issues and Policies

Additional books and e-books in this series can be found
on Nova's website under the Series tab.

AGRICULTURE ISSUES AND POLICIES

AGRICULTURAL ISSUES

POLICIES, CONSERVATION AND FARM PROGRAMS

MICHAEL BUTLIN
EDITOR

snova
New York

Library of Congress Cataloging-in-Publication Data

ISBN: 978-1-53615-474-0

Published by Nova Science Publishers, Inc. † New York

Contents

PREFACE

Agricultural Issues: Policies Conservation and Farm Programs is a compilation of government reports. The Agriculture appropriations bill, as described in chapter 1, funds all of USDA, excluding the U.S. Forest Service. It also funds the Food and Drug Administration (FDA) in the Department of Health and Human Services (HHS).

The Natural Resources Conservation Service (NRCS) and the Farm Service Agency (FSA) in the U.S. Department of Agriculture (USDA) currently administer 20 programs and subprograms that are directly or indirectly available to assist producers and landowners who wish to practice conservation on agricultural lands. These programs are discussed in chapter 2.

Chapter 3 provides background on the trade dispute that triggered the trade-aid package as well as the authority used by USDA to respond to the trade dispute with financial assistance and then describes the three components of the trade-aid package with details on their implementation.

For each crop year, the U.S. Department of Agriculture (USDA) makes billions of dollars in payments to agricultural producers for which being actively engaged in farming is a requirement. As described in chapter 4, the largest programs in terms of payments are the Price Loss Coverage program, which makes payments in years in which a crop's market price is less than a statutorily set price, and the Agriculture Risk Coverage program, which makes payments in years in which a crop's revenue is less than a revenue guarantee.

Chapter 5 describes cotton's special treatment, relative to other traditional farm program crops, in the 2014 farm bill.

The outlook for lower net farm income and relatively weak prices for most major program crops signals the likelihood of continued relatively lean times ahead. Chapter 6 incorporates USDA's August 30, 2018, farm income projections and its August 29, 2017, U.S. agricultural trade outlook update.

The 2014 Farm Act provides eligible U.S. farmers with new commodity supports in the Agriculture Risk Coverage (ARC), the Price Loss Coverage (PLC), and the Supplemental Coverage Option (SCO) programs. Chapter 7 provides an analysis of these programs with a focus on how various combinations of the programs impact producer revenue and its variability, producer well-being, and expected program costs.

Chapter 8 focuses on how specialty crops are covered under the federal crop insurance program.

The Animal and Plant Health Inspection Service (APHIS) of the U.S. Department of Agriculture (USDA) is the U.S. government authority tasked with regulating the import, transit, and release of regulated animals, animal products, veterinary biologics, plants, plant products, pests, organisms, soil, and genetically engineered organisms as reported in chapter 9.

Chapter 1 - The Agriculture appropriations bill funds the U.S. Department of Agriculture (USDA) except for the Forest Service. It also funds the Food and Drug Administration (FDA) and—in even-numbered fiscal years—the Commodity Futures Trading Commission (CFTC). Agriculture appropriations include both mandatory and discretionary spending. Discretionary amounts, though, are the primary focus during the bill's development. The largest discretionary spending items are the Special Supplemental Nutrition Program for Women, Infants, and Children (WIC); agricultural research; FDA; rural development; foreign food aid and trade; farm assistance programs; food safety inspection; conservation; and animal and plant health programs. The main mandatory spending items are the Supplemental Nutrition Assistance Program (SNAP), child nutrition, crop insurance, and the farm commodity and conservation programs. For FY2019, both the House and Senate Appropriations Committees reported agriculture bills (H.R. 5961, S. 2976) in May 2018. The Senate amended and passed its version as Division C of a four-bill minibus on August 1, 2018 (H.R. 6147). In the absence of a final appropriation, Congress enacted two continuing resolutions, but these expired on December 21, 2018 (P.L. 115- 245, Division C; and P.L. 115-298). Beginning December 22, 2018, a partial government shutdown (funding lapse) began limiting operations. Some "excepted" functions such as food safety and commodity grading can continue during the funding lapse. In the 116th Congress, the House passed H.R. 21, a six-bill omnibus with Agriculture as Division C. The Trump Administration requested $17 billion for discretionary-funded accounts within the jurisdiction of Agriculture appropriations, a reduction of $6.2 billion from FY2018 (-27%). Both the House-reported and Senate-passed bills generally reject most of the proposed reductions. The discretionary total of the House-reported bill is $23.23 billion, which would be $14 million less than enacted in FY2018 (-0.1%). The discretionary total of the Senate-passed bill is also $23.23 billion. However, the Senate bill's total would be $229 million more than enacted in FY2018 (+1%) on a comparable basis that excludes the CFTC. The Senate-passed bill would provide about $250 million more than the House-passed bill on a comparable basis. Among the primary differences between the bills and from FY2018: Both the House and Senate bills would increase FDA appropriations (+$308 million in the House bill; +$159 million in the Senate bill). Both bills increase appropriations for animal and plant health programs (+$16 million to +$19 million). The House bill would provide more base funding for rural water and waste disposal (+$81 million), but none of the extra that was provided separately in FY2018. The Senate bill would not change the base funding for rural water but continues $400 million of the extra funding. For rural broadband, both the House and Senate bills would continue a FY2018 pilot ($550 million in the House bill; $425 million in the Senate bill). The House bill would increase agricultural research (+$79 million), and the Senate bill would increase Agricultural Research Service programming (+98 million) but not fund any construction (-$141 million). Both bills provide less for WIC (-$175 million in the House bill, and -$25 million in the Senate bill), though the Senate bill has a larger rescission from prior-year WIC funds than does the House bill. The House bill would reduce base funding for the international Food for Peace program (-$100 million) and does not renew extra funding (-$116 million), while the Senate bill keeps it constant. The appropriations also carry mandatory spending—largely determined in separate authorizing laws—that would total $122 billion. Thus, the overall total of both bills is about $145 billion. Both bills contain policy provisions affecting disaster programs, rural definitions, industrial hemp, animal products, nutrition programs, dietary guidelines, CFTC, and tobacco products.

Chapter 2 - The Natural Resources Conservation Service (NRCS) and the Farm Service Agency (FSA) in the U.S. Department of Agriculture (USDA) currently administer 20 programs and subprograms that are directly or indirectly available to assist producers and landowners who wish to practice conservation on agricultural lands. The differences and number of these programs have created general confusion about the purpose, participation, and policies of the programs. While recent consolidation efforts removed some duplication, a large number of programs remain. The programs discussed in this chapter are as follows

- Agricultural Conservation Easement Program (ACEP)
- Agricultural Management Assistance (AMA)
- Conservation Operations (CO); Conservation Technical Assistance (CTA)
- Conservation Reserve Program (CRP)
- CRP—Conservation Reserve Enhancement Program (CREP)
- CRP—Farmable Wetland Program
- CRP—Grasslands
- Conservation Stewardship Program (CSP)
- Emergency Conservation Program (ECP)

- Emergency Forest Restoration Program (EFRP)
- Emergency Watershed Protection (EWP)
- Environmental Quality Incentives Program (EQIP)
- EQIP—Conservation Innovation Grants (CIG)
- Grassroots Source Water Protection Program
- Healthy Forests Reserve Program (HFRP)
- Regional Conservation Partnership Program (RCPP)
- Voluntary Public Access and Habitat Incentive Program
- Water Bank Program
- Watershed and Flood Prevention Operations
- Watershed Rehabilitation Program

This tabular presentation provides basic information covering each of the programs. In each case a brief program description is followed by information on major amendments in the Agricultural Act of 2014 (P.L. 113-79, 2014 farm bill), national scope and availability, states with the greatest participation, the backlog of applications or other measures of continuing interest, program funding authority, FY2018 funding, FY2019 Administration budget request, statutory authority, the authorization expiration date, and a link to the program's website.

Chapter 3 - In early 2018, the Trump Administration—citing concerns over national security and unfair trade practices—imposed increased tariffs on certain imported products in general and on U.S. imports from China in particular. Several of the affected foreign trading partners (including China) responded to the U.S. tariffs with their own retaliatory tariffs targeting various U.S. products, especially agricultural commodities. On July 24, 2018, Secretary of Agriculture Sonny Perdue announced that the U.S. Department of Agriculture (USDA) would be taking several temporary actions to assist farmers in response to trade damage from what the Administration has characterized as "unjustified retaliation." Specifically, the Secretary said that USDA would authorize up to $12 billion in financial assistance—referred to as a trade aid package—for certain agricultural commodities using Section 5 of the Commodity Credit Corporation (CCC) Charter Act (15 U.S.C. 714c). USDA intends for the trade aid package to provide short-term assistance in response to the ongoing trade disputes. However, the Secretary stated that there would not be further trade-related financial assistance beyond this $12 billion package. The aid package includes (1) a Market Facilitation Program (MFP) of direct payments (valued at up to $10 billion) to producers of soybeans, corn, cotton, sorghum, wheat, hogs, and dairy who are most affected by the trade retaliation (sweet cherries and almonds were added to this list in September); (2) a Food Purchase and Distribution Program to partially offset lost export sales of affected commodities ($1.2 billion); and (3) an Agricultural Trade Promotion (ATP) Program to expand foreign markets ($200 million). USDA's Farm Service Agency will administer the MFP by providing payments in two potential tranches: a first round announced on August 27, 2018, initially valued at $4.7 billion; and an equivalent-valued second round announced on December 17, 2018. However, producers need only sign up once for the MFP to be eligible for first and second payments. The sign-up period for soybeans, corn, cotton, sorghum, wheat, hogs, and dairy started September 4, 2018. The sign-up period for fresh sweet cherries and shelled almonds started on September 24. To be eligible, a producer must have an ownership share in the commodity, be actively engaged in farming, and be in compliance with adjusted gross income restrictions and conservation provisions. Eligible producers should apply after their harvest is complete. Initially, producers were given a deadline of January 15, 2019, to complete an application. However, USDA extended the deadline on January 8, to an indefinite time period due to the government shutdown. USDA used 2017 production data to estimate that approximately $9.6 billion would be distributed in MFP payments for corn, cotton, sorghum, soybeans, wheat, dairy, hogs, fresh sweet cherries, and shelled almonds, with over three-fourths ($7.3 billion) of MFP payments provided to soybean producers. MFP payments are capped on a per-person or per-legal-entity basis at a combined $125,000 for eligible crop commodities, a combined $125,000 for dairy production and hogs, and, separately, a combined $125,000 for fresh sweet cherries and shelled almonds. In addition to the MFP payments, the Administration announced a Food Purchase and Distribution Program that is to undertake $1.2 billion in government purchases of excess food supplies. USDA has targeted an initial 29 commodities for purchases and distribution through domestic nutrition assistance programs. Purchasing orders and distribution activities are to be adjusted based on the demand by the recipient food assistance programs geographically. The smallest piece of the trade aid package is an allocation of $200 million to the ATP to boost the trade promotion efforts at USDA's Foreign Agricultural Service, including foreign market development for affected agricultural products. USDA's use of its discretionary authority under the CCC Charter Act to make direct payments without further congressional action has historically been somewhat intermittent and limited in its scale. While the use of this authority is not without precedent, the scope and scale of this trade aid package has increased congressional and public interest. Furthermore, the significant variation in the announced MFP payment rates for affected commodities has elicited questions about equitable treatment among affected commodities. On September 13, USDA released a description of its MFP payment methodology, which is based strictly on the estimated direct trade "damage"—that is, export losses resulting from retaliatory tariffs. Indirect effects—such as the decline in market prices and resultant "lost value" for many of the affected commodities—were not included in the payment calculation.

Chapter 4 - This is an edited, reformatted and augmented version of U.S. Government Accountability Office, Publication No. GAO-18-384R, dated May 18, 2018.

Chapter 5 - The Bipartisan Budget Act of 2018 (P.L. 115-123; BBA), signed into law on February 9, 2018, included a provision—Section 60101(a)—which amended the 2014 farm bill (P.L. 113-79) to add seed cotton as a "covered commodity," thus making cotton eligible for the Price Loss Coverage (PLC) and Agricultural Risk Coverage (ARC) farm revenue support programs. The 2014 farm bill provides authority for farm programs for the 2014 through 2018 crop years. Under the 2014 farm bill as signed into law in February 2014, neither cotton nor its co-product, cottonseed, were eligible to participate in the newly created ARC and PLC programs. During the first three years of the 2014 farm bill (crop years

2014 through 2016), nearly $20 billion in support payments have been made to eligible program crops under ARC and PLC. Upland cotton had been eligible for U.S. Department of Agriculture (USDA) farm support programs since their origin in the 1930s. However, in response to a trade dispute with Brazil, the 2014 farm bill permanently removed upland cotton from eligibility for the ARC and PLC programs. Instead, the remaining cotton-specific support programs were supplemented by offering the U.S. cotton sector new transitional payments and, in 2015, a new shallow-loss revenue insurance product: the Stacked Income Protection Program (STAX). Since early 2014, U.S. cotton producers have been largely unsatisfied with their economic situation due to falling cotton prices and the perceived lack of revenue support offered under STAX. In early 2015, the U.S. cotton sector began advocating to be made eligible for payments under the ARC and PLC programs. Several Members of the House Agriculture Committee (including both its chairman and ranking member) were supportive of this effort and actively sought a legislative opportunity to reintegrate either cottonseed or seed cotton—the harvested but unginned cotton boll that includes both lint and cottonseed—back into the revenue support programs in the 2014 farm bill. After several legislative attempts, this effort was successful when Congress passed the BBA with a provision that amended the 2014 farm bill to add seed cotton as a "covered commodity," thus making it eligible for either ARC or PLC beginning with the 2018 crop. To participate in the seed cotton program, producers must first make three decisions specified in the new statute: what portion of their farm program generic base acres will be designated as cotton-specific base, whether to update historical cotton program yields (needed to determine the per-acre payment rate), and which program to participate in—ARC or PLC. The Congressional Budget Office (CBO) projects the cost of adding seed cotton as a covered commodity at nearly $3 billion over 10 years. However, accompanying budget offsets are projected to lower the net cost to $61 million. The policy change for cotton has potential implications for both domestic and international commodity markets. However, perhaps the most critical aspect of the new seed cotton policy is that, as part of current law, it becomes part of CBO's budget baseline used to write the next farm bill. As a result, it avoids the potential policy debate surrounding a search for budgetary offsets that would otherwise be needed to pay for the costs associated with adding a new farm program. This result could facilitate reaching an agreement among congressional negotiators working on the farm support programs in Title I of a new farm bill.

Chapter 6 - According to USDA's Economic Research Service (ERS), national net farm income—a key indicator of U.S. farm well-being—is forecast at $65.7 billion in 2018, down 13% from last year. This forecast does not include any payments under the trade aid package announced by USDA in July 2018. The forecast decline in 2018 net farm income is the result of lower cash returns—from both production activities (-0.1%) and government program payments (-17.4%)—and higher production expenses (up 3.3%). Net *farm* income is calculated on an accrual basis. Net *cash* income (calculated on a cash-flow basis) is also projected lower in 2018 (-12%) to $91.5 billion. The 2018 net farm income forecast is substantially below the 10-year average of $87 billion (in nominal dollars). It would be the lowest since 2016 in nominal and inflation-adjusted 2017 dollars. This is primarily the result of the outlook for continued weak prices for most major crops. U.S. farm income experienced a golden period during 2011 through 2014 due to strong commodity prices and robust agricultural exports. Most crops and livestock product prices remain significantly below the average for 2011-2014, when prices for many major commodities attained record or near-record highs. Net farm income is projected down 47% since its record high of $123.7 billion in 2013. Net cash income is projected down 32% from its 2013 high of $135.6 billion. Farm-sector production expenses have risen 1.6% over that same period, thus contributing to lower aggregate income totals. Government payments are projected down 17% from 2017 at $9.3 billion—due largely to projected lower payments in 2018 under the Price Loss Coverage (PLC) and Agricultural Risk Coverage (ARC) revenue support programs for major field crops (projected at $3.5 billion, down 50% from $7.0 billion in 2017). Since 2008, U.S. agricultural exports have accounted for a 20% share of U.S. farm and manufactured or processed agricultural sales. In 2018 agricultural exports are forecast to be flat, at $144.5 billion, due largely to abundant supplies in international markets and strong competition from major foreign competitors. The ongoing U.S.-China trade dispute is expected to shift trade patterns, but the aggregate value of U.S. exports is expected to hold steady. In spite of the lower farm income outlook, farm wealth is projected up 1.2% from 2017 to $3.0 trillion. Farm asset values reflect farm investors' and lenders' expectations about long-term profitability of farm sector investments. Farmland values are projected up 1.9% in 2018 following a 2.3% rise in 2017. Because they comprise such a significant portion of the U.S. farm sector's asset base (81%), change in farmland values is a critical barometer of the farm sector's financial performance. At the farm household level, average farm household incomes have been well above average U.S. household incomes since the late 1990s. In 2016 (the last year with comparable data), the average farm household income (including off-farm income sources) of $117,918 was about 42% higher than the average U.S. household income of $83,143. The outlook for lower net farm income and relatively weak prices for most major program crops signals the likelihood of continued relatively lean times ahead. With record soybean and near-record corn crops predicted by USDA, the U.S. agricultural sector's well-being remains dependent on continued growth in domestic and foreign demand to sustain prices at current modest levels. In addition to commodity prices, the financial picture for the agricultural sector as a whole heading into 2019 will hinge on both domestic and international macroeconomic factors, including interest rates, economic growth, and consumer demand. This chapter incorporates USDA's August 30, 2018, farm income projections and its August 29, 2017, U.S. agricultural trade outlook update.

Chapter 7 - The 2014 Farm Act provides eligible U.S. farmers with new commodity supports in the Agriculture Risk Coverage (ARC), the Price Loss Coverage (PLC), and the Supplemental Coverage Option (SCO) programs. These programs help producers with revenue losses generally not covered by traditional crop insurance policies. Interactions, both among these programs and between these programs and the Federal Crop Insurance (FCI) program, determine the nature and magnitude of support available to producers. This chapter provides an analysis of these programs with a focus on how various combinations of the programs impact producer revenue and its variability, producer well-being, and expected program costs. The report finds that these programs' effectiveness are influenced by historical prices, expected prices, and FCI coverage rates. High historic crop prices combined with low expected prices since the enactment of the 2014 Farm Act led to higher enrollment of producers in the ARC program in 2015 relative to that in the SCO program.

Chapter 8 - The federal crop insurance program offers subsidized crop insurance policies to farmers. Farmers can purchase policies that pay indemnities when their yields or revenues fall below guaranteed levels. While the majority of federal crop insurance policies cover yield or revenue losses, the program also offers policies with other types of guarantees, such as index policies that trigger an indemnity payment based on weather conditions. The Federal Crop Insurance Corporation (FCIC), a government corporation within the U.S. Department of Agriculture (USDA), pays part of the premium—about 63%, on average—across the federal crop insurance portfolio during crop year 2017, while policy holders—farmers and ranchers—pay the balance. Private insurance companies, known as Approved Insurance Providers (AIPs), deliver the policies in return for administrative and operating subsidies from FCIC. AIPs also share underwriting risk with FCIC through a mutually negotiated Standard Reinsurance Agreement. The USDA Risk Management Agency (RMA) administers the federal crop insurance program. The federal crop insurance program primarily covers traditional field crops (such as wheat, corn, and soybeans) that are supported by USDA's revenue-support programs. Unlike these traditional crops, *specialty crops*—defined in statute as "fruits and vegetables, tree nuts, dried fruits, and horticulture and nursery crops (including floriculture)" (7 U.S.C. §1621 note) — have not been a major part of federal crop insurance support. Specialty crops are also generally not eligible for USDA's revenue-support programs. USDA estimates that the statutory definition of *specialty crops* covers more than 300 agricultural commodities, including fresh and processed fruits and vegetables, tree nuts, nursery plants (trees, shrubs, and flowering plants).

herbs, spices, coffee, tea, honey, and maple syrup. Legislative changes, coupled with ongoing administrative efforts by USDA, have expanded federal crop insurance coverage for specialty crops, and they now account for a small but growing number of federal crop insurance policies bought by farmers. Among the issues Congress may consider if it seeks to further expand coverage for specialty crops are data collection and price discovery for commodities not sold on exchanges (such as most fruits and vegetables), coverage of quality losses, and the effect of ad hoc payments on the demand for crop insurance. Federal crop insurance policies currently cover around 38 specialty crop categories, which include roughly 80 types of fruits, vegetables, tree nuts, and nursery crops. Over the past few decades, total specialty crop insured liabilities rose from nearly $1 billion in 1989 to nearly $18.5 billion in 2017. In 2017, federal crop insurance policies covered about 9 million acres of specialty crops, around 800,000 bee colonies, about 100,000 tons of raisins, and roughly 60,000 fruit and coffee trees. Across all specialty crops, coverage and the premium subsidy paid by the federal government may vary depending on the crop. Moreover, for many specialty crops, crop-specific insurance policies are not available. Currently, about one-half of all U.S. specialty crop acres are covered by federal crop insurance policies. Some specialty crops may be covered under a Whole Farm Revenue Protection (WFRP) insurance policy. WFRP is designed to fill in coverage gaps for producers of uninsured crops that lack individual policy coverage and for producers marketing to local, farm-identity preserved, or direct markets. The average premium subsidy rate for WRFP was about 70% in 2017. Federal crop insurance for specialty crops and WFRP together accounted for about 17% of the entire federal crop insurance portfolio, as measured by liability, during crop year 2017.

Chapter 9 - The Animal and Plant Health Inspection Service (APHIS) of the U.S. Department of Agriculture (USDA) is the U.S. government authority tasked with regulating the import, transit, and release of regulated animals, animal products, veterinary biologics, plants, plant products, pests, organisms, soil, and genetically engineered organisms. APHIS provides scientific authorities in trade partner countries and U.S. importers with animal and plant health import regulations. APHIS requires U.S. importers to obtain animal or plant health import permits, which verify that the items being imported meet U.S. import standards. Animal and plant health import permits certify that imports follow U.S. regulations, World Trade Organization (WTO) guidelines, and/or trading partner specific requirements. These import permits are a part of broader agreements between the United States and its trading partners within the WTO on established sanitary and phytosanitary (SPS) measures. These measures aim to protect against diseases, pests, toxins, and other contaminants. The House and Senate Agricultural Appropriations Committees appropriate funds that allow APHIS to carry out a range of activities, including those involved in issuing import permits. From FY2014 to FY2018, discretionary appropriations for APHIS have averaged nearly $900 million. About 85% of the APHIS budget is allocated to the "Safeguarding and Emergency Preparedness/Response" mission area, which includes the administration of health import permits and other efforts to prevent imports of pests and diseases into the United States. APHIS's authority over agricultural imports is largely provided by the Animal Health Protection Act (7 U.S.C. §§8301 *et seq.*), the Plant Protection Act (7 U.S.C. §§7701 *et seq.*), and the Public Health Security and Bioterrorism Preparedness and Response Act of 2002 (7 U.S.C. §§8401). These laws authorize APHIS to administer animal and plant health import permits and conduct agricultural import inspections. APHIS works with other federal agencies, such as the Department of Homeland Security's Customs and Border Protection (CBP), to conduct animal and plant health monitoring programs and to determine if new pest or disease management programs are needed. In addition, Congress directs the Food and Drug Administration, the Food Safety and Inspection Service, and state-level Departments of Agriculture to participate in inspecting many products regulated by APHIS. APHIS and CBP personnel inspect shipments of imported agricultural products and certify that the required import health permits and SPS certificates accompany each shipment. One of the major flagship programs that APHIS and CBP administer together is the Agricultural Quarantine Inspection (AQI) program, in which APHIS and CBP technical staff work to ensure that the required animal or plant health permits, sanitary certificates (for animal products), and phytosanitary certificates (for plant products) accompany each shipment. APHIS transfers funds to CBP to conduct AQI activities. The ongoing congressional commitment to preventing plant and animal disease and pests from entering the United States through agricultural imports is evident in annual appropriations Congress provides for APHIS. Congress has directed APHIS to monitor pests and diseases and has assigned APHIS to oversee SPS activities in some free trade agreements. Moreover, legislation introduced in the 115th Congress sought to address invasive species (e.g., Areawide Integrated Pest Management, H.R. 5411) and would have directed CBP to enforce APHIS regulations to deter smuggling of plants and animals into the United States.

In: Agricultural Issues
Editor: Michael Butlin
ISBN: 978-1-53615-474-0

Chapter 1

AGRICULTURE AND RELATED AGENCIES: FY2019 APPROPRIATIONS (UPDATED)*

Jim Monke

SUMMARY

The Agriculture appropriations bill funds the U.S. Department of Agriculture (USDA) except for the Forest Service. It also funds the Food and Drug Administration (FDA) and—in even-numbered fiscal years—the Commodity Futures Trading Commission (CFTC).

Agriculture appropriations include both mandatory and discretionary spending. Discretionary amounts, though, are the primary focus during the bill's development. The largest discretionary spending items are the Special Supplemental Nutrition Program for Women, Infants, and Children (WIC); agricultural research; FDA; rural development; foreign food aid and trade; farm assistance programs; food safety inspection; conservation; and animal and plant health programs. The main mandatory spending items are the Supplemental Nutrition Assistance Program

* This is an edited, reformatted and augmented version of Congressional Research Service Publication No. R45230, Updated January 8, 2019.

(SNAP), child nutrition, crop insurance, and the farm commodity and conservation programs.

For FY2019, both the House and Senate Appropriations Committees reported agriculture bills (H.R. 5961, S. 2976) in May 2018. The Senate amended and passed its version as Division C of a four-bill minibus on August 1, 2018 (H.R. 6147). In the absence of a final appropriation, Congress enacted two continuing resolutions, but these expired on December 21, 2018 (P.L. 115- 245, Division C; and P.L. 115-298). Beginning December 22, 2018, a partial government shutdown (funding lapse) began limiting operations. Some "excepted" functions such as food safety and commodity grading can continue during the funding lapse. In the 116th Congress, the House passed H.R. 21, a six-bill omnibus with Agriculture as Division C.

The Trump Administration requested $17 billion for discretionary-funded accounts within the jurisdiction of Agriculture appropriations, a reduction of $6.2 billion from FY2018 (-27%). Both the House-reported and Senate-passed bills generally reject most of the proposed reductions.

The discretionary total of the House-reported bill is $23.23 billion, which would be $14 million less than enacted in FY2018 (-0.1%). The discretionary total of the Senate-passed bill is also $23.23 billion. However, the Senate bill's total would be $229 million more than enacted in FY2018 (+1%) on a comparable basis that excludes the CFTC. The Senate-passed bill would provide about $250 million more than the House-passed bill on a comparable basis.

Among the primary differences between the bills and from FY2018: Both the House and Senate bills would increase FDA appropriations (+$308 million in the House bill; +$159 million in the Senate bill). Both bills increase appropriations for animal and plant health programs (+$16 million to +$19 million). The House bill would provide more base funding for rural water and waste disposal (+$81 million), but none of the extra that was provided separately in FY2018. The Senate bill would not change the base funding for rural water but continues $400 million of the extra funding. For rural broadband, both the House and Senate bills would continue a FY2018 pilot ($550 million in the House bill; $425 million in the Senate bill). The House bill would increase agricultural research (+$79 million), and the Senate bill would increase Agricultural Research Service programming (+98 million) but not fund any construction (-$141 million).

Both bills provide less for WIC (-$175 million in the House bill, and -$25 million in the Senate bill), though the Senate bill has a larger rescission from prior-year WIC funds than does the House bill. The House bill would reduce base funding for the international Food for Peace program (-$100 million) and does not renew extra funding (-$116 million), while the Senate bill keeps it constant.

The appropriations also carry mandatory spending—largely determined in separate authorizing laws—that would total $122 billion. Thus, the overall total of both bills is about $145 billion.

Both bills contain policy provisions affecting disaster programs, rural definitions, industrial hemp, animal products, nutrition programs, dietary guidelines, CFTC, and tobacco products.

STATUS OF FY2019 AGRICULTURE APPROPRIATIONS

In the absence of a final appropriation, FY2019 began under two continuing resolutions that lasted until December 21, 2018 (*Table 1*). Beginning December 22 a partial government shutdown began limiting operations. Some "excepted" functions can continue during the funding lapse. On January 3, 2019 the House passed an omnibus bill (H.R. 21) to reopen the government, using the text of a Senate-passed bill from the 115th Congress.

In 2018, both the House and Senate Appropriations Committees reported Agriculture appropriations bills for FY2019 (H.R. 5961 on May 16, 2018, and S. 2976 on May 24, 2018). The Senate amended and passed its version as Division C of a four-bill minibus (H.R. 6147 on August 1, 2018). See *Figure 1* and *Appendix B* for a more complete picture of the timeline.

Higher discretionary budget caps in the Bipartisan Budget Act of 2018 (P.L. 115-123) have facilitated development of the appropriation. The discretionary total of the House-reported bill was $23.23 billion, $14 million less than enacted in FY2018 (-0.1%; *Table 2*). The discretionary total of the Senate-passed bill was also $23.23 billion, $229 million more than enacted in FY2018 (–1%) on a comparable basis that excludes the

Table 1. Status of FY2019 Agriculture Appropriations

House Action			Senate Action				
Subcmte.	Cmte.	Floor	Subcmte.	Cmte.	Floor	Continuing Resolutions	Final Appropriation
115th Congress							
5/9/2018	5/16/2018	—	5/22/2018	5/24/2018	8/1/2018	P.L. 115-245,	—
Draft[a]	H.R. 5961		Draft	S. 2976	H.R. 6147	Division C,	
Voice vote	H.Rept. 115-706		Voice vote	S.Rept. 115-259	Division C Vote 92-6	until 12/7/2018 P.L. 115-298	
	Vote 31-20			Vote 31-0		until 12/21/18	
116th Congress							
—	—	1/3/2019	—	—	—	—	—
		H.R. 21					
		Vote of 241-190					

Source: CRS, compiled from the Legislative Information System.

[a] The House subcommittee draft is at https://docs.house.gov/meetings/AP/AP01/ 20180509/108287/ BILLS115HR-SC-AP-FY2019-Agriculture-SubcommitteeDraft.pdf, and the committee report draft is at https://docs.house. gov/meetings/AP/AP00/20180516/108312/HRPT-115-HR-FY2019-Agriculture.pdf.

Table 2. Agriculture and Related Agencies Appropriations, by Title, FY20 1 8-FY2019 (budget authority in millions of dollars)

	FY2018		FY2019			Change FY2018 to FY2019...	
Title of Agriculture Appropriations Act	*P.L. 115-141*	*Admin. Request*	*House H.R. 5961*	*Senate H.R. 6147*	*Enacted*	*House*	*Senate*
I. Agricultural Programs: Discretionary	5,622.8	4,752.9	5,704.5	5,587.8		+81.7	-35.1
Mandatory	*1,344.0*	*1,374.0*	*1,374.0*	*1,374.0*	—	*+30.0*	*+30.0*
Subtotal	[illegible]	[illegible]	[illegible]	[illegible]		[illegible]	[illegible]
II. Farm Production and Conservation Programs	2,735.6	2,197.9	2,760.3	2,728.0	—	+24.7	-7.6
Mandatory	*23,198.3*	*24,097.5*	*24,097.5*	*24,097.5*	—	*+899.2*	*+899.2*
Subtotal	*25,933.9*	*26,295.4*	*26,857.8*	*26,825.5*	—	*+923.9*	*+891.6*
III. Rural Development	3,000.9a	2,022.5	3,078.5[a]	3,000.9[a]	—	+77.6	+0.0
IV. Domestic Food Programs: Discretionary	6,709.8	5,990.1	6,528.5	6,696.1	—	-181.2	-13.7
Mandatory	*98,209.6*	*96,342.3*	*96,344.8*	*96,344.8*	—	*-1,864.9*	*-1,864.9*
Subtotal	*104,919.4*	*102,332.3*	*102,873.3*	*103,040.9*	—	*-2,046.1*	*-1,878.5*
V. Foreign Assistance	2,021.0[a]	205.0	1,925.8	2,152.3	—	-95.2	+131.4
VI. Related Agencies:							
Food and Drug Administration	2,811.9[a]	3,183.7	3,119.6	2,970.9	—	+307.7	+159.0
Commodity Futures Trading Commission	249.0	281.5	255.0	[281.5][b]	—	+6.0	[+32.5][b]

Table 2. (Continued)

	FY2018		FY2019			Change FY2018 to FY2019...	
Title of Agriculture Appropriations Act	*P.L. 115-141*	*Admin. Request*	*House H.R. 5961*	*Senate H.R. 6147*	*Enacted*	*House*	*Senate*
VII. General Provisions:							
CHIMPS and rescissions[c]	-801.0	-1,197.8	-315.0	-395.0	—	+486.0	+406.0
Other appropriations	1,378.1	0.0	579.5	889.5	—	-798.6	-488.6
Scorekeeping adjustments[d]	-481.0	-409.0	-404.0	-404.0	—	+77.0	+77.0
Discretionary: Senate basis w/o CFTC	[22,998.0]	16,745.2	[22,977.8]	23,226.5	—	-20.2	+228.5
Discretionary: House basis w/ CFTC	23,247.0	17,026.7	23,232.8	[23,508.0][b]	—	-14.2	+261.0[b]
Mandatory	*122,752.0*	*121,813.8*	*121,816.3*	*121,816.3*	—	-935.7	-935.7
Total: Senate basis w/o CFTC	*[145,750.0]*	*138,559.0*	*[144,794.0]*	*145,042.8*	—	-955.9	-707.2
Total: House basis w/ CFTC	*145,999.0*	*138,840.5*	*145,049.0*	[145,324.3][b]	—	-949.9	-674.7[b]

Source: CRS, using appropriations text and reports and unpublished CBO tables.

Notes: Amounts are nominal discretionary budget authority in millions of dollars unless labeled otherwise.

[a] Excludes amounts in supplemental appropriations acts. Bracketed amounts are not in the official totals due to differing House-Senate jurisdiction for CFTC.

[b] See Division B of H.R. 6 147.

[c] Changes in Mandatory Program Spending (CHIMPS) are reductions made to mandatory programs via appropriations. Rescissions are permanent cancellations of previously provided budget authority.

[d] "Scorekeeping adjustments" are not necessarily appropriated but are part of the official CBO accounting.

Commodity Futures Trading Commission (CFTC).[1] The Senate-passed bill would have provided about $250 million more than the House-reported bill on a comparable basis.

The appropriations also carry mandatory spending—though that is largely determined in separate authorizing laws—that would total nearly $122 billion. Thus, the overall total of each of the bills is about $145 billion.

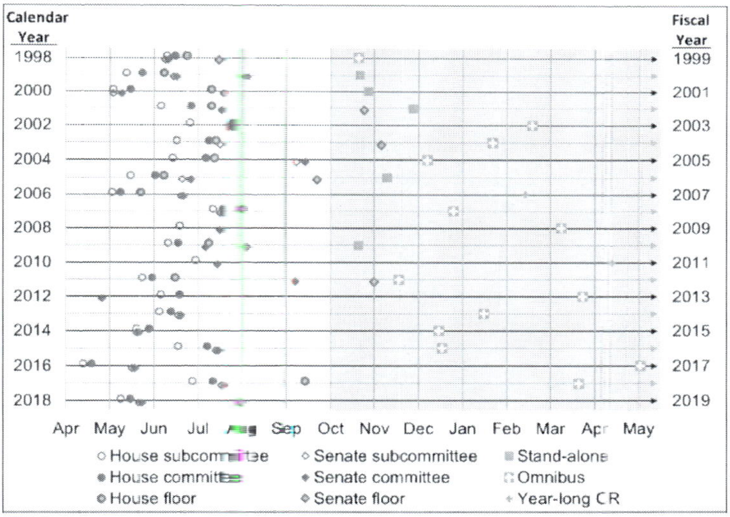

Source: CRS.

Figure 1. Timeline of Action on Agriculture Appropriations, FY1999-FY2019.

SCOPE OF AGRICULTURE APPROPRIATIONS

The Agriculture appropriations bill—formally known as the Agriculture, Rural Development, Food and Drug Administration, and Related Agencies Appropriations Act—funds all of USDA, excluding the

---

[1] Jurisdiction for CFTC appropriations differs between the chambers. Since FY2008, CFTC is marked up in the Agriculture Subcommittee of the House Appropriations Committee and in the Financial Services and General Government Subcommittee of the Senate Appropriations Committee. The enacted CFTC appropriation is carried in the Agriculture bill in even-numbered fiscal years and in the Financial Services bill in odd-numbered fiscal years.

U.S. Forest Service. It also funds the Food and Drug Administration (FDA) in the Department of Health and Human Services (HHS). In even-numbered fiscal years, the act carries CFTC funding under a practice for handling House-Senate jurisdictional differences. Jurisdiction is with the House and Senate Committees on Appropriations and their respective Subcommittees on Agriculture, Rural Development, Food and Drug Administration, and Related Agencies. The bill includes mandatory and discretionary spending, but the discretionary amounts are the primary focus during the bill's development. The scope of the bill is shown in *Figure 2*. The federal budget process treats discretionary and mandatory spending differently:[2]

- Discretionary spending is controlled by annual appropriations acts and receives most of the attention during the appropriations process. The annual budget resolution[3] process sets spending limits for discretionary appropriations. Agency operations (salaries and expenses) and many grant programs are discretionary.
- Mandatory spending[4]—though carried in the appropriation and usually advanced unchanged—is controlled by budget rules during the authorization process.[5] Spending for so-called entitlement programs is set in laws such as the 2014 farm bill[6] and 2010 child nutrition reauthorizations.[7]

---

[2] See CRS Report R44582, Overview of Funding Mechanisms in the Federal Budget Process, and Selected Examples.

[3] See CRS Report R42388, The Congressional Appropriations Process: An Introduction.

[4] Mandatory spending creates funding stability and consistency compared to annual discretionary appropriations. In agriculture, it originated with farm commodity programs that had uncertain outlays due to weather and markets.

[5] See CRS Report 98-560, Baselines and Scorekeeping in the Federal Budget Process.

[6] P.L. 113-79. See CRS In Focus IF10783, Farm Bill Primer: Budget Issues.

[7] P.L. 111-296. See CRS Report R44373, Tracking the Next Child Nutrition Reauthorization: An Overview.

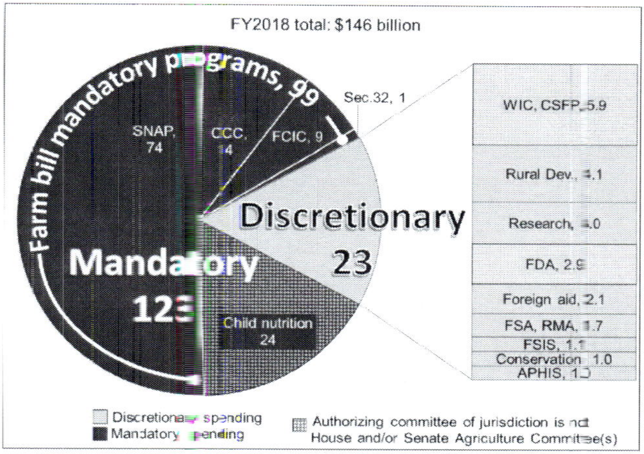

Source: CRS. Does not show some agencies under $0.5 billion. Includes General Provisions with agencies.

Notes: SNAP = Supplemental Nutrition Assistance Program; CCC = Commodity Credit Corporation; FCIC = Federal Crop Insurance Corporation; Section 32 = Funds for Strengthening Markets, Income and Supply; WIC = Special Supplemental Nutrition Program for Women, Infants, and Children; CSFP = Commodity Supplemental Food Program; FDA = Food and Drug Administration; FSA = Farm Service Agency; RMA = Risk Management Agency; FSIS = Food Safety and Inspection Service; APHIS = Animal and Plant Health Inspection Service.

Figure 2. Scope of Agriculture and Related Agencies Appropriations (FY2018 budget authority in billions of dollars)

In the FY2018 Agriculture appropriations, discretionary appropriations were 16% ($23.3 billion) of the $146 billion total. Mandatory spending carried in the act comprised $123 billion, about 84% of the total. About $99 billion of the $123 billion mandatory amount could be attributed to programs in the 2014 farm bill (*Figure 2*). Some programs are not in the authorizing jurisdiction of the House or Senate Agriculture Committees.

Within the discretionary total, the largest discretionary spending items are the Special Supplemental Nutrition Program for Women, Infants, and Children (WIC); rural development; agricultural research; FDA; foreign food aid and trade; farm assistance program salaries and loans; food safety inspection; conservation; and animal and plant health programs (*Figure 2*).

The main mandatory spending items are the Supplemental Nutrition Assistance Program (SNAP, and other food and nutrition act programs), child nutrition (school lunch and related programs), crop insurance, and farm commodity and conservation programs that are paid through USDA's Commodity Credit Corporation (CCC). SNAP is referred to as an "appropriated entitlement" and requires an annual appropriation.[8]

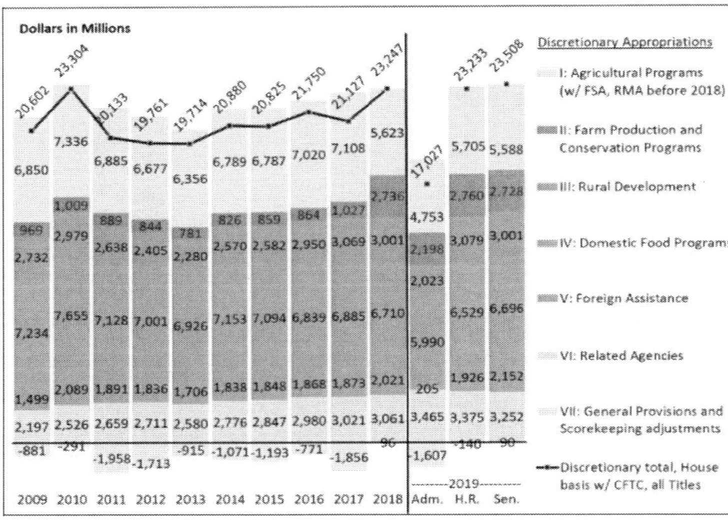

Source: CRS.

Note: For FY2019, Adm. is the Administration's request, H.R. is committee-reported H.R. 5961 and Sen. is Senate-passed H.R. 6147. Includes CFTC in Related Agencies in all years.

Figure 3. Discretionary Agriculture Appropriations, by Title, FY2009-FY2019.

Amounts for the nutrition program are based on projected spending needs. In contrast, the CCC operates on a line of credit. The annual appropriation provides funding to reimburse the Treasury for the use of this line of credit.

---

[8] See CRS Report RS20129, Entitlements and Appropriated Entitlements in the Federal Budget Process.

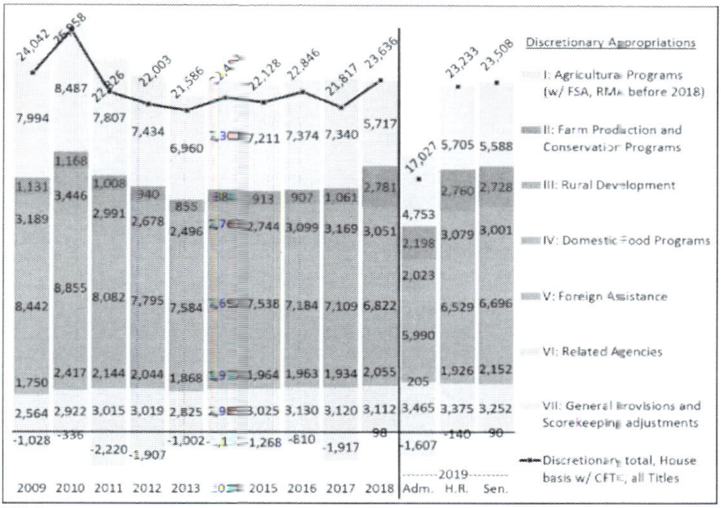

Source: CRS.

Note: For FY2019, Adm. is the Administration's request, H.R. is committee-reported H.R. 5961 and Sen. is Senate-passed H.R. 6147. Includes CFTC in Related Agencies in all years. Budget authority is millions of dollars adjusted to FY2019 using the gross domestic product price deflator.

Figure 4. Inflation-Adjusted Agriculture Appropriations, FY2009-FY2019.

ACTION ON FY2019 APPROPRIATIONS

Administration's Budget Request

The Trump Administration released its FY2019 budget request on February 12, 2018,[9] one week later than the usual first Monday in February. USDA concurrently released its more detailed budget summary and justification,[10] as did the FDA,[11] and the independent agencies of the

[9] OMB, *FY2019 Budget of the U.S. Government*, Appendix, https://www.whitehouse.gov/omb/appendix.

[10] USDA, *FY2019 USDA Budget Summary*; and USDA, *2019 Budget Explanatory Notes*, http://www.obpa.usda.gov

[11] FDA, *FY2019 FDA Justification of Estimates for Appropriations*, https://www.hhs.gov/about/budget/index.html.

CFTC[12] and the Farm Credit Administration (FCA).[13] The Administration also highlighted some of the proposed reductions and eliminations separately.[14]

From these documents, the congressional appropriations committees evaluated the request and began to consider their own bills in the spring of 2018. For accounts in the jurisdiction of the Agriculture appropriations bill, the Administration's budget proposed $17.4 billion, a 25% reduction from FY2018 (*Table 2*, *Figure 3*).

The timing of the Administration's budget request for FY2019 preceded Congress enacting the final, omnibus FY2018 appropriation in March 2018. Therefore, amounts in the FY2018 column of the Administration's budget documents are based on FY2017 and the continuing resolution and are not reliable indicators of the enacted FY2018 levels that came later.

Discretionary Budget Caps and Subcommittee Allocations

Budget enforcement for appropriations has both procedural and statutory elements.

The procedural elements are associated with the budget resolution and are enforced through points of order. Typically, each chamber's full Appropriations Committee receives a top-line procedural limit on discretionary budget authority, referred to as a "302(a)" allocation, from the Budget Committee via an annual budget resolution passed by each chamber. The Appropriations Committees then each in turn subdivide their allocation among their subcommittees, referred to as the "302(b)" allocations.[15]

The statutory elements impose limits on discretionary spending in FY2012-FY2021 and are enforced through discretionary budget caps and

---

[12] CFTC, *FY2019 CFTC Budget Request*, https://www.cftc.gov/sites/default/files/idc/groups/public/documents/file/cftcbudget2019.pdf.

[13] FCA, *FY2019 FCA Proposed Budget*, https://www.fca.gov/Download/BudgetFY2019.pdf.

[14] OMB, *FY2019 Budget of the U.S. Government*, *Major Savings and Reforms.*

[15] CRS Report R42388, The Congressional Appropriations Process: An Introduction.

sequestration (2 U.S.C. 901(c)).[16] The Budget Control Act of 2011 (BCA, P.L. 112-25) set discretionary budget caps through FY2021 as a way of reducing federal spending.[1] Bipartisan Budget Acts (BBAs) in 2013, 2015, and 2018 (P.L. 113-67, P.L. 114-74, and P.L. 115-123, respectively) have avoided sequestration on discretionary spending—with the exception of FY2013—by raising those caps (*Appendix A*).[18]

In February 2018, the BBA raised the FY2019 cap on nondefense discretionary spending by $68 billion and the cap on defense spending by $85 billion.[19] It also provided language to execute (or "deem") those higher caps for the appropriations process without following the usual procedures for an FY2019 budget resolution.[20]

Under these higher caps and authorities, the House and Senate Appropriations Committees proceeded to mark up the FY2019 appropriations bills. The Agriculture appropriations bills are receiving roughly the same subcommittee allocation ("302(b)" allocation) in each chamber that was used to complete the FY2018 appropriation under the BBA. For FY2019, the subcommittee allocations for agriculture appropriations are

- House: $23.273 billion (H.Rept. 115-710, May 23, 2018), including the CFTC.
- Senate: $23.235 billion (S.Rept. 115-260, May 24, 2018), excluding the CFTC.

---

[16] CRS Report R42972, Sequestration as a Budget Enforcement Process: Frequently Asked Questions.

[17] CRS Report R43411, The Budget Control Act of 2011: Legislative Changes to the Law and Their Budgetary Effects.

[18] Looking to the future, though, the budget caps for FY2020 and FY2021 remain unchanged from the original BCA in 2011. They are less than the revised caps for FY2019. Unless the caps change, this implies less future spending.

[19] CRS Insight IN10861, Discretionary Spending Levels Under the Bipartisan Budget Act of 2018.

[20] CRS Report R45157, The Bipartisan Budget Act of 2018 and an FY2019 Budget Resolution.

House Action

115th Congress (2018)

The House Agriculture Appropriations Subcommittee marked up its FY2019 bill on May 9, 2018, by voice vote.[21] On May 16, 2018, the full Appropriations Committee passed and reported an amended bill (H.R. 5961, H.Rept. 115-706) by a vote of 31-20 (Table 1, Figure 1).

The $23.23 billion discretionary total in the House-reported FY2019 Agriculture appropriation would be $14 million less than enacted in FY2018 (-0.1%; Table 2, Figure 3). Generally speaking, the House-reported bill does not include most of the reductions proposed by the Administration and continues the trend of appropriations from prior years.

Table 3 provides details at the agency level. The primary changes from FY2018 at the agency level that comprise the relatively flat $14 million overall decrease in the House-reported bill would do the following:

- Increase base FDA appropriations by $308 million (+11%). However, it does not continue to separate FDA funding for the opioid crisis that was in the General Provisions title of the FY2018 appropriation (-$94 million).
- Increase agricultural research (+$79 million, +2.6%) by raising appropriations for the Agricultural Research Service and National Institute of Agriculture.
- Increase the Animal and Plant Health Inspection Service (APHIS) by $16 million (+1.7%).
- Increase the Rural Utility Service by $82 million (+13%), mostly in the base funding for rural water and waste disposal programs. However, it does not continue separate funding for rural water that was in the General Provisions title last year (-$500 million) or for telemedicine for the opioid crisis (-$20 million). It reduces

---

[21] The House subcommittee draft is at https://docs.house.gov/meetings/AP/AP01/20180509/108287/BILLS-115HRSC-AP-FY2019-Agriculture-SubcommitteeDraft.pdf, and the draft of the committee report is at https://docs.house.gov/ meetings/AP/AP00/20180516/108312/HRPT-115-HR-FY2019-Agriculture.pdf.

separate funding for a broadband pilot program while continuing to fund some of it through the General Provisions title (-$50 million).

- Decrease discretionary appropriations for domestic nutrition assistance programs by reducing WIC by $175 million (-2.8%) and the commodity assistance programs by $15 million (-4.7%). However, the reduction scored in the General Provisions title by rescinding WIC carryover balances is smaller in FY2019, retaining more budget authority (+$500 million).
- Decrease the base funding in international food assistance Food for Peace grants by $100 million (-6.2%) and not continue the extra funding that was in the General Provisions title of the FY2018 appropriations (-$116 million).

In addition to discretionary spending, the appropriations also carry funding for mandatory spending—largely determined in separate authorizing laws—that would total $121.82 billion, about $936 million less than in FY2018 because of automatic changes in economic conditions and entitlement enrollment rather than any change from congressional action. Thus, the overall total of the House-reported bill is about $145 billion.

116th Congress (2019)

As the 116th Congress began in 2019 under a funding gap (see the heading "Government Shutdown"), the House passed a six-bill full-year omnibus appropriation on January 3, 2019 (H.R. 21). The Agriculture appropriations portion is Division C of H.R. 21.

With little exception for agriculture, it is essentially identical to the Senate-passed Agriculture appropriations bill from the 115th Congress (Division C of H.R. 6147; see under the heading "Senate Action").

- One funding difference is that the rescission from WIC is $501 million in Section 724 of H.R. 21, compared to $400 million in the same section of H.R. 6147.

- H.R. 21 does not contemplate enactment of the 2018 farm bill (P.L. 115-334) in December 2018 in various provisions that prescribe how funding should be implemented in light of 2014 farm bill provisions. For example, H.R. 21 would fund the "Assistant to the Secretary for Rural Development" under the Office of the Secretary rather than recognizing Section 12407 of the 2018 farm bill that reestablishes the Senate-confirmed position "Under Secretary of Agriculture for Rural Development."
- H.R. 21 follows Senate jurisdiction for CFTC by placing its funding with the Financial Services appropriation rather than in the Agriculture appropriations, as would be the custom for appropriations in House bills.

While the House text of Division C of H.R. 21 is essentially the Senate-passed Agriculture appropriations bill from the 115th Congress, the Senate would need to reintroduce and consider the bill for the 116th Congress.

Senate Action

115th Congress (2018)

The Senate Agriculture Appropriations Subcommittee marked up a FY2019 bill on May 22, 2018, by voice vote. On May 24, 2018, the Appropriations Committee passed and reported an amended bill (S. 2976, S.Rept. 115-259) by a vote of 31-0. On August 1, 2018, the Senate passed a four-bill minibus (H.R. 6147) by a vote of 92-6, with agriculture as Division C (*Table 1, Figure 1*).

The discretionary total of the Senate-passed bill is also $23.23 billion. However, the Senate bill's total would be $229 million more than enacted in FY2018 (+1%) on a comparable basis that excludes the CFTC, since the latter was part of the enacted FY2018 appropriation. The Senate-passed bill would provide about $250 million more than the House-reported bill would on a comparable basis computed by subtracting CFTC from the House bill.

Table 3 provides details at the agency level. The primary changes from FY2018 at the agency level that comprise the Senate-passed bill's overall $228 million increase are the following:

- Increase base FDA appropriations by $159 million (+6%). Like the House bill, it does not continue separate FDA funding for the opioid crisis that was in the General Provisions title of the FY2018 appropriation (-$94 million).
- Increase APHIS by $19 million (+1.9%), slightly more than the House bill.
- Decrease discretionary appropriations for domestic nutrition assistance programs by reducing WIC by $25 million (-0.4%). This is a smaller reduction than in the House bill. In the rescissions in the General Provisions title, the reduction is smaller than was rescinded in FY2019 (+$400 million). Nonetheless, the Senate bill's rescission is greater than in the House bill.
- Increase international nutrition assistance by raising the base funding for Food for Peace grants by $116 million (+7.2%). The extra funding that was in the General Provisions title of the FY2018 appropriations is not continued (-$116 million).
- Decrease the four agricultural research agencies by $44 million (-1.4%), mostly by providing no funding for ARS building and facilities (-$141 million), while increasing ARS salaries and expenses (+$98 million).
- Decrease the extra funding for rural development compared to the amount provided in the FY2018 General Provisions (-$100 million for rural water, -$475 million for broadband). Base funding for rural development is unchanged overall.

The Senate-passed bill's mandatory spending is virtually identical to that of the House-reported bill ($121.82 billion). Its overall total of discretionary and mandatory appropriations is $145 billion.

Continuing Resolutions

In the absence of a final FY2019 Agriculture appropriation, Congress enacted two continuing resolutions (CRs) to continue government operations through December 21, 2018 (P.L. 115-245, Division C, from 10/1/2018 through 12/7/2018, and P.L. 115-298 through 12/21/2018).[22] On December 19, 2018, the Senate passed another CR (H.R. 695) that would have continued temporary funding through February 8, 2019, but the House-passed amendment to that bill on December 20 added Homeland Security funding for construction of physical barriers at borders and supplemental appropriations for natural disasters that the Senate did not accept.

When the last CR (P.L. 115-298) expired without further enacted budget authority, a partial government shutdown began on December 22. Some "excepted" functions such as food safety and commodity grading can continue during the funding lapse, as explained in "Government Shutdown."

In general, a CR continues the funding rates and conditions that were in the previous year's appropriation.[23] The Office of Management and Budget (OMB) may prorate funding to the agencies on an annualized basis for the duration of the CR through a process known as apportionment.[24] For the 81 days (22%) of FY2019 through December 21, 2018, the CRs

- continued the terms of the FY2018 Agriculture appropriations act (Section 101) and exclude the FY2018 change in mandatory program spending (CHIMP) on the Biomass Crop Assistance Program, which is not authorized for FY2019; and

---

[22] CRS Report R42647, Continuing Resolutions: Overview of Components and Recent Practices.

[23] CRS Report RL34700, Interim Continuing Resolutions (CRs): Potential Impacts on Agency Operations.

[24] For example, if a CR lasts for three months, then 3/12 of the previous fiscal year amount may be apportioned to limit agency spending. See OMB, "Apportionment of the Continuing Resolution(s) for Fiscal Year 2019," September 28, 2018, https://www.whitehouse.gov/wp-content/uploads/2017/11/bulletin_no_18_05.pdf.

- provided sufficient funding to maintain mandatory program levels, including for nutrition programs (Section 110)—this is standard language for recent years' CRs but is additionally important for SNAP, because some authorizations in the 2014 farm bill began expiring after FY2018.[25]

CRs may adjust prior-year amounts through anomalies or make specific administrative changes. Five anomalies specifically applied to the agriculture appropriation during both CRs:

1. Child Nutrition Programs: apportionment for a summer foods program sufficient to allow it to be operational by May 2019 (Section 114).
2. Rural Utilities Service: allows a loan authorization level for the Rural Water and Waste Disposal program of $4.141 billion (Section 115).
3. Commodity Credit Corporation (CCC): allows CCC to receive its transfer about a month earlier than usual, prior to a final report and audit. Many farm bill payments to farmers are due in October 2018 in addition to USDA's plan to make supplemental payments under a trade assistance program.[26] Without the anomaly, CCC might exhaust its $30 billion line of credit (Section 116).
4. Agricultural Research Service: an additional $42 million for operations and maintenance at the National Bio and Agro-Defense Facility (NBAF) being built in Manhattan, Kansas, and being transferred to USDA from the Department of Homeland Security (Section 117).
5. Department of Homeland Security (DHS): DHS may transfer up to $15 million to USDA to support NBAF operations (Section 125).

---

[25] CRS In Focus IF10989, Expiration of the 2014 Farm Bill: Some Potential Implications.

[26] CRS Report R45310, *Farm Policy: USDA's Trade Aid Package*.

Government Shutdown

When an appropriation (or CR) expires and no further budget authority has been provided, a funding gap exists and may cause operations to cease at affected agencies. In general, the Antideficiency Act (31 U.S.C. 1341 et seq.) prohibits federal agencies from obligating funds before an appropriations measure has been enacted. Exceptions may allow for certain activities to continue, such as for law enforcement, protection of human life or property, and activities funded by other means such as carryover funds or user fees. Programs that are funded by other authorities—such as entitlements or the mandatory programs in the farm bill—may also be affected if the program is executed using personnel whose salaries are funded by discretionary appropriations that are affected by the funding gap.[27]

For FY2019, a funding gap began on December 22, 2018, for agencies within the jurisdiction of 7 of the 12 appropriations bills, including the Agriculture appropriations bill. The last time that the government experienced a multiweek shutdown was in October 2013. Prior to that, the next previous multiday shutdown was in FY1996, though Agriculture appropriations were not affected then because a stand-alone full-year Agriculture appropriation had been enacted.[28]

In general, a shutdown implies the furlough of certain personnel and curtailment of affected agency activities and services. Agencies make their own determinations about activities and personnel that are "excepted" from furlough and publish their intentions in "contingency plans" that are supervised by OMB.[29] For agencies in the Agriculture appropriations

---

[27] CRS Report RL34680, *Shutdown of the Federal Government: Causes, Processes, and Effects*.

[28] CRS Report RS20348, *Federal Funding Gaps: A Brief Overview*.

[29] OMB, "Agency Contingency Plans," https://www.whitehouse.gov/omb/information-for-agencies/agencycontingency-plans.

jurisdiction, shutdown or contingency plans were published for USDA,[30] FDA,[31] and the CFTC.[32]

USDA initially estimated on December 23 that 56% of its employees were excepted from furlough in the agencies that are funded by Agriculture appropriations (all of USDA except the Forest Service), which amounts to 35,004 staff being excepted out of 62,288.[33] The number of personnel varies by agency, and it may change as a shutdown continues as funding availability changes and as new circumstances arise.

For example, nearly 90% of the FSIS and AMS staff (8,434 and 3,944 staff, respectively) were initially retained to continue food safety inspections of meat and poultry at processing plants and to continue commodity grading and inspection services for commerce. About 60% of the APHIS was excepted (4,769 staff) to continue preclearance inspection for transportation between Hawaii and Puerto Rico and the mainland and to carry out quarantine and certification for imports and exports. Resources in research laboratories and facilities that could be damaged by inattention was protected by excepting 18% (1,116 staff) of ARS. All of NRCS was initially excepted by using mandatory funding (9,342 staff).[34] The Farm Service Agency initially excepted 6,427 staff (61% out of 10,479) through December 28 but then closed county offices, thereby lowering the number of excepted employees.[35]

---

[30] USDA, "U.S. Department of Agriculture Lapse in Funding Plans," various agency-level documents, December 2018 https://www.usda.gov/shutdownplans.

[31] HHS, "FY2019 HHS Contingency Staffing Plan for Operations in the Absence of Enacted Annual Agriculture and Interior Appropriations," https://www.hhs.gov/about/budget/fy-2019-hhs-contingency-staffing-plan/index.html.

[32] CFTC, "CFTC Plan for Lapse in Appropriations," December 18, 2018, https://www.cftc.gov/sites/default/files/2018- 12/CFTCPlanLapseAppropriations121818.pdf.

[33] CRS analysis of USDA in "Summary of Activities to be Continued in the Event of a Government Shutdown," December 23, 2018, https://www.usda.gov/sites/default/ files/documents/usda-shutdown-plan-summary-2.pdf.

[34] USDA Natural Resources Conservation Service, "NRCS Is Open for Business," https://www.nrcs.usda.gov/wps/ portal/nrcs/detail/national/home/?cid=NRCSEPRD1434827 (accessed January 2, 2019).

[35] Farm Service Agency, "Government Shutdown Information," https://www.fsa.usda.gov/help/shutdowninfo, accessed January 2, 2019.

While a number of selected USDA functions continue under the shutdown, many others have ceased operations. Examples of USDA functions that are not being performed by furloughed employees include data collection and analysis that inform the commodity markets, development of regulations to implement the new farm bill that was enacted in December 2018, completing the Administration's "trade aid" payments, processing and funding farm loans and guarantees, rural development loan and grant programs (rural housing, community facilities and rural water, rural business and broadband), agricultural research programs and grants, and many international assistance programs. The duration of the shutdown could expand the number of USDA activities affected.

Most of USDA's Food and Nutrition Service (FNS) programs, whether mandatory or discretionary, rely on funding provided in appropriations acts. FNS program operations during a government shutdown vary based on the different programs' available resources, determined by factors such as contingency or carryover funds and terms of the expired appropriations acts, as well as USDA's decisionmaking. Beginning in late December 2018, FNS released program-specific memoranda to states and program operators describing the status of different nutrition assistance programs during the current funding lapse.[36]

FDA initially retained 59% of its employees in duty status (10,344 staff out of 17,397) based on a combination of factors including carryover funding, the need to safeguard human life, and the protection of property.[37] CFTC excepted 9% of its employees (61 staff out of 673) to address risks that could pose a threat to the functioning of the stock market and commodity markets and that could affect the safety of human life or the protection of property.[38]

---

[36] FNS, "Information on Program Operations in Light of Lapse in Fiscal Year 2019 Appropriations for the Food and Nutrition Service," https://www.fns.usda.gov/FY19Lapse.

[37] HHS, "FY2019 Contingency Staffing Plan," December 21, 2018, https://www.hhs.gov/sites/default/files/fy-2019-hhs-lapse-contingency-plan-narrative-december-ag-interior.pdf.

[38] CFTC, "CFTC Plan for Lapse in Appropriations."

Table 3. Agriculture and Related Agencies Appropriations, by Agency, FY2016-FY2019 (budget authority in millions of dollars)

Agency or Major Program	FY2016 P.L. 114-113	FY2017 P.L. 115-31	FY2018 P.L. 115-141	FY2019 Admin. Request	FY2019 House H.R. 5961	FY2019 Senate H.R. 6147	FY2019 Enacted	Change FY18 to FY19 House	Change FY18 to FY19 Senate
Title I. Agricultural Programs									
Departmental Administration	373.2	403.9	396.0	369.8	387.8	389.5	—	-8.1	-6.5
[illegible]									
Agricultural Research Service	1,355.9	1,269.8	1,343.4	1,019.0	1,395.9	[illegible]	—	[illegible]	[illegible]
National Institute of Food and Agriculture	1,326.5	1,367.9	1,407.8	1,257.7	1,452.6	1,423.2	—	+44.8	+15.4
National Agricultural Statistics Service	168.4	171.2	191.7	165.0	173.7	174.8	—	-18.0	-17.0
Economic Research Service	85.4	86.8	86.8	45.0	86.8	86.8	—	+0.0	+0.0
Under Secretary	0.9	0.9	0.8	0.8	0.8	0.8	—	+0.0	+0.0
Marketing and Regulatory Programs									
Animal and Plant Health Inspection Service	897.6	949.4	985.1	742.0	1,001.5	1,003.7	—	+16.5	+18.6
Agricultural Marketing Service[a]	82.5	86.2	152.8	119.7	154.3	157.1	—	+1.5	+4.3
Section 32 (M)	1,303.0	1,322.0	1,344.0	1,374.0	1,374.0	1,374.0	—	+30.0	+30.0
Grain Inspection, Packers, Stockyards Admin.[a]	43.1	43.5	moved into Agricultural Marketing Service[a]						
Under Secretary	0.9	0.9	0.9	0.8	0.9	0.9	—	+0.0	+0.0
Food Safety									
Food Safety and Inspection Service	1,014.9	1,032.1	1,056.8	1,032.3	1,049.3	1,049.3	—	-7.5	-7.5
Under Secretary	0.8	0.8	0.8	0.8	0.8	0.8	—	+0.0	+0.0
Farm and Commodity Programs[b]									
Farm Service Agency[b]	1,595.1	1,624.0	moved to Title II: Farm Production and Conservation[b]						
FSA Farm Loans: Loan Authority[c]	*6,402.1*	*8,002.6*	moved to Title II: Farm Production and Conservation[b]						

Table 3. (Continued)

Agency or Major Program	FY2016 P.L. 114-113	FY2017 P.L. 115-31	FY2018 P.L. 115-141	FY2019 Admin. Request	FY2019 House H.R. 5961	FY2019 Senate H.R. 6147	FY2019 Enacted	Change FY18 to FY19 House	Change FY18 to FY19 Senate
Risk Management Agency Salaries and Exp.	74.8	74.8	moved to Title II: Farm Production and Conservation[a]						
Federal Crop Insurance Corporation (M)	7,858.0	8,667.0	moved to Title II: Farm Production and Conservation[a]						
Commodity Credit Corporation (M)	6,871.1	21,290.7	moved to Title II: Farm Production and Conservation[a]						
Under Secretary	0.9	0.9	moved to Title II: Farm Production and Conservation[a]						
Subtotal									
Discretionary	**7,020.3**	**7,107.7**	**5,622.8**	**4,752.9**	**5,704.5**	**5,593.8**	—	**+81.7**	**-29.1**
Mandatory (M)	*16,032.6*	*31,280.2*	*1,344.0*	*1,374.0*	*1,374.0*	*1,374.0*	—	*+30.0*	*+30.0*
Subtotal	*23,052.9*	*38,387.9*	*6,966.8*	*6,126.9*	*7,078.5*	*6,967.8*	—	*+111.7*	*+0.9*
Title II. Farm Production and Conservation[a]									
Business Center	—	—	1.0	196.4	115.4	1.0	—	+114.4	+0.0
Farm Service Agency[b]	moved from Title I[a]		1,625.2	1,294.2	1,518.8	1,622.7	—	-106.4	-2.6
FSA Farm Loans: Loan Authority[c]	moved from Title I[a]		*8,005.6*	*7,617.7*	*7,987.7*	*8,017.7*	—	*-17.9*	*+12.1*
Risk Management Agency Salaries and Exp.	moved from Title I[a]		74.8	37.9	75.4	74.8	—	+0.6	+0.0
Federal Crop Insurance Corporation (M)	moved from Title I[a]		8,913.0	8,687.0	8,687.0	8,687.0	—	-226.0	-226.0
Commodity Credit Corporation (M)	moved from Title I[a]		14,284.8	15,410.0	15,410.0	15,410.0	—	+1,125.2	+1,125.2
Conservation Operations	850.9	864.5	874.1	669.0	890.3	879.1	—	+16.2	+5.0
Watershed and Flood Prevention	—	150.0	150.0	—	150.0	150.0	—	+0.0	+0.0
Watershed Rehabilitation Program	12.0	12.0	10.0	—	10.0	—	—	+0.0	-10.0
Under Secretary	0.9	0.9	0.9	0.9	0.9	0.9	—	+0.0	+0.0
Subtotal									
Discretionary	**863.8**	**1,027.4**	**2,735.6**	**2,197.9**	**2,760.3**	**2,728.0**	—	**+24.7**	**-7.6**
Mandatory (M)	moved from Title I I[a]		*23,198.3*	*24,097.5*	*24,097.5*	*24,097.5*	—	*+899.2*	*+899.2*
Subtotal	moved from Title I I[a]		*25,933.9*	*26,295.4*	*26,857.8*	*26,825.5*	—	*+923.9*	*+891.6*

Agency or Major Program	FY2016 P.L. 114-113	FY2017 P.L. 115-31	FY2018 P.L. 115-141	FY2019 Admin. Request	FY2019 House H.R. 5961	FY2019 Senate H.R. 6147	FY2019 Enacted	Change FY18 to FY19 House	Change FY18 to FY19 Senate
Title III. Rural Development									
Salaries and Expenses (including transfers)[a]	682.9	675.8	680.8	585.9	686.8	682.8	—	+6.0	+2.0
Rural Housing Service	1,616.4	1,654.9	1,582.4	1,351.4	1,582.5	1,585.2	—	+0.2	+2.8
RHS Loan Authority[c]	*27,496.8*	*28,083.4*	*28,390.1*	*27,760.0*	*28,345.5*	*28,590.1*	—	*-44.6*	*+200.0*
Rural Business-Cooperative Service[a]	90.5	97.7	109.5	0.0	99.5	104.2	—	-10.0	-5.4
RBCS Loan Authority[c]	*979.3*	*988.4*	*991.2*	*0.0*	*1,027.5*	*991.2*	—	*+36.3*	*+0.0*
Rural Utilities Service	559.3	639.9	628.1[f]	85.2	709.6[f]	628.7[f]	—	+81.5	+0.6
RUS Loan Authority[c]	*8,210.6*	*8,217.0*	*8,219.9*	*7,413.1*	*8,419.9*	*8,219.9*	—	*+200.0*	*+0.0*
Under Secretary	0.9	0.9	moved to Departmental Administration as an Assistant to the Secretary[d]						
Subtotal, Discretionary	**2,950.0**	**3,069.2**	**3,000.9[f]**	**2,022.5**	**3,078.5[f]**	**3,000.9[f]**	—	**+77.6**	**+0.0**
Subtotal, RD Loan Authority[c]	*36,686.7*	*37,288.9*	*37,601.2*	*35,173.1*	*37,792.9*	*37,801.2*	—	*+191.6*	*+200.0*
Title IV. Domestic Food Programs									
Child Nutrition Programs (M)	22,149.7	22,794.0	24,254.1	23,146.9	23,183.5	23,184.0	—	-1,070.6	-1,070.1
WIC Program	6,350.0	6,350.0	6,175.0	5,750.0	6,000.0	6,150.0	—	-175.0	-25.0
SNAP, Food and Nutrition Act Programs (M)	80,849.4	78,480.7	74,013.5	73,218.3	73,219.3	73,219.3	—	-794.2	-794.2
Commodity Assistance Programs	296.2	315.1	322.1	55.5	306.9	322.1	—	-15.2	+0.0
Nutrition Programs Administration	150.8	170.7	153.8	160.8	162.8	164.7	—	+9.0	+10.8
Under Secretary	0.8	0.8	0.8	0.8	0.8	0.8	—	+0.0	+0.0
Subtotal									
Discretionary	**6,838.9**	**6,884.7**	**6,709.8**	**5,990.1**	**6,528.5**	**6,696.1**	—	**-181.2**	**-13.7**
Mandatory (M)	*102,958.1*	*101,226.7*	*98,209.6*	*96,342.3*	*96,344.8*	*96,344.8*	—	*-1,864.9*	*-1,864.9*
Subtotal	*109,797.0*	*108,111.3*	*104,919.4*	*102,332.3*	*102,873.3*	*103,040.9*	—	*-2,046.1*	*-1,878.5*

Table 3. (Continued)

Agency or Major Program	FY2016	FY2017	FY2018	FY2019				Change FY18 to FY19	
	P.L. 114-113	P.L. 115-31	P.L. 115-141	Admin. Request	House H.R. 5961	Senate H.R. 6147	Enacted	House	Senate
Title V. Foreign Assistance									
Foreign Agricultural Service	191.6	196.6	199.7	193.1	204.1	212.2	—	+4.5	+12.6
Food for Peace Title II, and admin. expenses	1,468.5[f]	1,466.1[f]	1,600.1[f]	0.1	1,500.1	1,716.1	—	-100.0	+116.0
McGovern-Dole Food for Education	201.6	201.6	207.6	0.0	207.6	210.3	—	+0.0	+2.6
CCC Export Loan Salaries	6.7	8.5	8.8	7.1	9.2	8.8	—	+0.3	+0.0
Office of Codex Alimentarius	—	—	3.8	3.8	3.8	4.0	—	+0.0	+0.2
Under Secretary	—	—	0.9	0.9	0.9	0.9	—	+0.0	+0.0
Subtotal	**1,868.5**	**1,872.9**	**2,021.0**	**205.0**	**1,925.8**	**2,152.3**	—	-95.2	+131.4
Title VI. Related Agencies									
Food and Drug Administration	2,729.6	2,771.2	2,811.9	3,183.7	3,119.6	2,970.9	—	+307.7	+159.0
Commodity Futures Trading Commission[g]	250.0	[250.0]	249.0	281.5	255.0	[281.5][h]	—	+6.0	[+32.5][h]
Subtotal	**2,979.6**	**[3,021.2]**	**3,060.9**	**3,465.2**	**3,374.6**	[3,252.4][h]	—	+313.7	[+191.5][h]
Title VII. General Provisions									
Reductions in Mandatory Programs[i]									
a. Environmental Quality Incentives Program	-209.0	-179.0	—	-136.3	—	—	—	+0.0	+0.0
b. Watershed Rehabilitation Program	-68.0	-54.0	—	-46.2	—	—	—	+0.0	+0.0
c. Fresh Fruit and Vegetable Program	-125.0	-125.0	—	-125.0	—	—	—	+0.0	+0.0
d. Biorefinery Assistance Program	-19.0	-20.0	—	—	—	—	—	+0.0	+0.0
e. Biomass Crop Assistance Program	-20.0	-20.0	-21.0	—	—	—	—	+21.0	+21.0
f. Cushion of Credit (Rural Development)	-179.0	-132.0	—	-225.0	-50.0	—	—	-50.0	+0.0
g. Section 32	-216.0	-231.0	—	-342.0	—	—	—	+0.0	+0.0
h. Other CHIMPS and mandatory rescissions	+5.0	+17.0	+20.0	-36.4	+35.0	+5.0	—	+15.0	-15.0

Agency or Major Program	FY2016 P.L. 114-113	FY2017 P.L. 115-31	FY2018 P.L. 115-141	FY2019 Admin. Request	FY2019 House H.R. 5961	FY2019 Senate H.R. 6147	FY2019 Enacted	Change FY18 to FY19 House	Change FY18 to FY19 Senate
Subtotal, CHIMPS	**-831.0**	**-744.0**	**-1.0**	**-910.8**	**-15.0**	**+5.0**	—	-14.0	+6.0
Rescissions (discretionary)	**-34.0**	**-854.0**	**-800.0**	-287.0	-300.0	-400.0	—	+500.0	+400.0
Other appropriations									
a. Disaster/emergency programs	273.0	234.8	—	—	—	—	—	+0.0	+0.0
b. Water and Waste Water	—	—	+500.0	—	—	+400.0	—	-500.0	-100.0
c. Broadband pilot	—	—	+600.0	—	+550.0	+425.0	—	-50.0	-175.0
d. Opioid Enforcement and Surveillance	—	—	+94.0	—	—	—	—	-94.0	-94.0
e. Food for Peace	250.0	131.0	116.0					116.0	116.0
f. Other appropriations	22.1	[illegible]	[illegible]		[illegible]	[illegible]		[illegible]	[illegible]
Subtotal, Other appropriations	**556.1**	**472.2**	**1,378.1**	—	**579.5**	**889.5**	—	-798.6	-488.6
Total, General Provisions	**-308.9**	**-1,125.3**	**577.1**	**-1,197.8**	**264.5**	**494.5**	—	-312.6	-82.6
Scorekeeping Adjustments[a]									
Disaster declaration in this bill	-130.0	-206.1	—	—	—	—	—	+0.0	+0.0
Other scorekeeping adjustments	-332.0	-524.0	-481.0	-409.0	-404.0	-404.0	—	+77.0	+77.0
Subtotal, Scorekeeping adjustments	**-462.0**	**-730.1**	**-481.0**	**-409.0**	**-404.0**	**-404.0**	—	+77.0	+77.0
Totals									
Discretionary: Senate basis w/o CFTC	[21,500.0]	20,877.0	[22,998.0]	16745.2	[22,977.8]	23,226.5	—	-20.2	+228.5
Discretionary: House basis w/ CFTC	21,750.0	[21,127.0]	23,247.0	17026.7	23,232.8	[23,508.0][b]	—	-14.2	[+261.0][b]
Mandatory (M)	*118,990.7*	*132,506.9*	*122,752.0*	121813.8	121,816.3	121,816.3	—	-935.7	-935.7
Total: Senate basis w/o CFTC	*[140,490.7]*	*153,383.9*	*[145,750.0]*	138559.0	[144,794.0]	145,042.8	—	-955.9	-707.2
Total: House basis w/ CFTC	*140,740.7*	*[illegible]*	*[illegible]*	138840.5	145,049.0	[illegible]		-919.9	-671.7[b]

Source: CRS, using appropriations text and report tables, and unpublished CBO tables.

Notes: Amounts are nominal discretionary budget authority in millions of dollars unless labeled otherwise. (M) indicates that the account is mandatory authority (or primarily mandatory authority). Excludes amounts in supplemental appropriations acts. Bracketed amounts are not in the official totals due to differing House-Senate jurisdiction for CFTC but are shown for comparison.

Table 3. (Continued)

[a] Row headings reflect recent USDA reorganization. The Farm Service Agency and Risk Management Agency were moved from Title I to Title II, as was the Commodity Credit Corporation and Federal Crop Insurance Corporation in mandatory spending. Grain Inspection, Packers, and Stockyards Administration was moved into the Agricultural Marketing Service.

[b] Includes regular FSA salaries and expenses, plus transfers for farm loan program salaries and administrative expenses. Also includes farm loan program loan subsidy, State Mediation Grants, Dairy Indemnity Program (mandatory funding), and Grassroots Source Water Protection Program. Does not include appropriations to the Foreign Agricultural Service for export loans and P.L. 480 administration that are transferred to FSA.

[c] Loan authority is the amount of loans that can be made or guaranteed with a loan subsidy. This amount is not added in the budget authority subtotals or totals.

[d] Includes Rural Development salaries and expenses and transfers from the three rural development agencies for salaries and expenses. Amounts for the agencies thus reflect program funds for loans and grants.

[e] Amounts for the Rural Business-Cooperative Service (RBCS) are before the rescission in the Cushion of Credit account, unlike in Appropriations Committee tables. The rescission is included with the CHIMPS as classified by CBO, which allows the RBCS subtotal to remain positive.

[f] Excludes a portion of the other appropriations that are provided separately in General Provisions.

[g] Jurisdiction for CFTC is in the House Agriculture Appropriations Subcommittee and the Senate Financial Services Appropriations Subcommittee. After FY2008, CFTC is carried in enacted Agriculture appropriations in even-numbered fiscal years in House Agriculture markup but not in Senate Agriculture markup. Bracketed amounts are not in the official totals due to differing House-Senate jurisdiction for CFTC but are shown for comparison.

[h] See Division B of H.R. 6147.

[i] Includes reductions (limitations and rescissions) to mandatory programs that may also be known as Changes in Mandatory Program Spending (CHIMPS).

[j] Includes $206 million appropriated for the Emergency Conservation Program (ECP) and Emergency Watershed Program (EWP) in the FY2017 second CR (P.L. 114- 254, §185) that were offset as emergency spending. Another $29 million for ECP was included in the final appropriation (§753).

[k] "Scorekeeping adjustments" are not necessarily appropriated items and may not be shown in appropriations committee tables but are part of the official CBO score (accounting) of the bill. They predominantly include "negative subsidies" in loan program accounts and adjustments for disaster designations in the bill.

Policy-Related Provisions

In addition to setting budgetary amounts, the Agriculture appropriations bill has also been a vehicle for policy-related provisions that direct how the executive branch should carry out the appropriation. These provisions may have the force of law if they are included in the text of the appropriation, usually in the General Provisions, but their effect is generally limited to the current fiscal year. Sometimes, the provisions amend the U.S. Code and have long-standing effects.

Table 4. Selected Policy Provisions in FY2019 Agriculture Appropriations

House-reported H.R. 5961	Senate-passed H.R. 6147, Division C
Disaster payments. In general, prohibits the use of Section 32, clause (3), to reestablish farmers' purchasing power by making payments to farmers. However, allows an exception to use up to $350 million of carryover for this purpose. (§715)	Similar to House provision. (§715)
Rural definition. Defines *rural* for the water and waste disposal guaranteed loan program as a city, town, or unincorporated area with no more than 20,000 people. (§728) Directs that eligibility for rural development programs shall not include incarcerated prison populations. (§744	Similar to House provisions. (§§728, 744)
Persistent poverty counties. Requires that at least 10% of the funds in certain rural development programs shall be allocated to persistent poverty counties, defined as any county that has had 20% or more of its population living in poverty over the past 30 years. (§750)	No comparable provision.
Income verification. Provides USDA access to Social Security and Internal Revenue Service data to verify the income of individuals participating in certain rural housing programs. (§747)	Similar to House provision. (§748)

Table 4. (Continued)

House-reported H.R. 5961	Senate-passed H.R. 6147, Division C
American steel. Prohibits funding for the rural water, wastewater, waste disposal, and solid waste management projects unless all of the iron and steel products are produced in the United States. (§743)	Similar to House provision. (§742)
No comparable provision.	Agriculture risk coverage (ARC) pilot. Directs USDA to conduct a pilot program to make payments for the ARC program that addresses disparities in the calculation of yields among counties. (§747)
No comparable provision.	Industrial hemp. Prohibits funding in contravention of the 2014 farm bill provision for research on industrial hemp or the transportation, processing, sale, or use of industrial hemp or seeds that are grown in accordance with the farm bill provision. (§729)
No comparable provision.	National Bio- and Agro-defense Facility (NBAF). Authorizes ARS and APHIS to appoint up to 50 employees annually in FY2019-2025 for NBAF at a rate of pay that exceeds the General Schedule. (§738)
No comparable provision.	Horse slaughter. Prohibits USDA from conducting horse slaughter inspection. (§758)
Animal Welfare Act. Prohibits funding to issue or renew licenses to class B dealers who sell dogs and cats for use in research, experiments, teaching, or testing. (§741)	No comparable provision.
Lab-grown meat. Directs USDA to regulate products that are made from cells of amenable species of livestock that are grown under controlled conditions for human food to prevent adulteration and misbranding. Applies to FY2018 and thereafter. (§736)	No comparable provision.
Poultry from China. Prohibits funds to buy raw or processed poultry products from China for the National School Lunch Program, School Breakfast Program, Child and Adult Care Food Program, and Summer Food Service Program. (§748) Prohibits funding to finalize the proposed rule, "Eligibility of the People's Republic of China to Export to the United	No comparable provisions.

House-reported H.R. 5961	Senate-passed H.R. 6147, Division C
States Poultry Products," until USDA verifies certain conditions. (§751)	
School meal prices. Exempts certain school food authorities (those without a negative balance in their school food service accounts as of December 31, 2018) from paid meal equity requirements in school year 2019-2020. (§757)	No comparable provision.
Vegetables in school breakfast. Prohibits funding to implement or enforce the portion of a School Breakfast Program regulation that limits substituting fruits with certain vegetables. (§767)	No comparable provision.
SNAP retailer standards. Prohibits funding to administer the "variety requirements" in the final rule "Enhancing Retailer Standards in SNAP" until the Secretary amends the definition of *variety* to increase the number of items that qualify as acceptable varieties in each staple food category. (§727)	Similar to House provision. (§728)
SNAP disclosure. Exempts from Freedom of Information Act disclosure SNAP transaction data that contain information specific to a store, store location, person, or other entity. (§768)	No comparable provision.
Human embryos. Prohibits FDA from using funds to accept any investigational new drug application for "research in which a human embryo is intentionally created or modified to include a heritable genetic modification." (§732)	Similar to House provision. (§733)
Partially hydrogenated oils. Directs FDA that no food containing partially hydrogenated oils shall be considered to be adulterated on that basis so long as it enters interstate commerce before June 18, 2018. (§740)	Similar to House provision. (§736)
Sodium. Prevents FDA from developing regulations or issuing final guidance for population-wide sodium reduction actions until a dietary reference intake report is completed. (§752)	Similar to House provision. (§750)
Grapes for wine. Prohibits FDA funds to enforce the final rule "Standards for the Growing, Harvesting, Packing and Holding of Produce for Human Consumption" with respect to grape varietals used solely for wine. (§755)	Similar to House provision. (§753)

Table 4. (Continued)

House-reported H.R. 5961	Senate-passed H.R. 6147, Division C
Dietary guidelines. Directs USDA to submit a report to Congress within 180 days about the process used to establish the 2020-2025 Dietary Guidelines for Americans, including the recommendations from the National Academies of Science report, "Redesigning the Process for Establishing the Dietary Guidelines for Americans." (§763)	No comparable provision.
Added sugars. Prohibits funds to administer the final rule "Food Labeling: Revision of the Nutrition and Supplement Facts Labels" to the extent that it requires added sugars labeling for single-ingredient foods to which sugars or sweeteners are not added (e.g., honey or maple syrup). (§764)	Similar to House provision. (§768)
Genetically engineered salmon. Directs that disclosure requirements for genetically engineered salmon or finfish follow the National Bioengineered Food Disclosure Standard and subsequent USDA regulations. (§766)	Prohibits FDA from allowing genetically engineered salmon into interstate commerce until it publishes final labeling guidelines. (§740)
CFTC statutes. Amends the Commodity Exchange Act to address the initial margin requirements for certain swap transactions. (§738) Further amends the act by directing CFTC to consider cost-benefit analysis in its regulations (using language from a House-passed bill, H.R. 238). (§760)	No comparable provision.
Cigar regulation. Prohibits funds to administer the FDA final "deeming rule" from May 10, 2016, with respect to traditional large and premium cigars. (§769) Prohibits funds to regulate cigars and pipe tobacco differently than April 25, 2014, the date that such tobacco products were proposed to be deemed subject to certain requirements. (§770)	No comparable provision.
Tobacco products. Amends the Federal Food, Drug, and Cosmetic Act concerning the registration of substantially equivalent tobacco products. (§771)	No comparable provision.

House-reported H.R. 5961	Senate-passed H.R. 6147, Division C
Vapor products. Directs the Secretary of Health and Human Services to issue a notice of proposed rulemaking, within 21 months, establishing a product standard for vapor products and to promulgate a final rule within 36 months. (§772) Deems a vapor product to be misbranded if it is advertised in a non-adult publication, specifies terms for distributing vapor products in other than direct face-to-face sales, directs the Secretary of Health and Human Services to promulgate regulations requiring labeling of vapor products, and imposes registration requirements on vapor product retailers. (§773) Amends the Pro-Children Act of 1994 to extend the prohibition on smoking in certain facilities that provide children's services to vapor products. (§774) Amends the Federal Food, Drug, and Cosmetic Act to add age verification requirements for remote sales of vapor products. (§775) Directs the Secretary of Health and Human Services to submit a report to Congress within 180 days about the Youth Vapor Product Education, Prevention, and Enforcement Program. (§776)	No comparable provision.
Electronic nicotine delivery system (ENDS). Directs the FDA commissioner to conduct a study and submit a report to Congress within 180 days on preventing the use of ENDS by youth. (§777)	No comparable provision.
No comparable provision.	Prevents certain changes from occurring under implementation of the Grain Standards Act. (§759)
No comparable provision.	Provides for technical assistance to respond to volcanic eruptions. (§760)
No comparable provision.	Directs USDA and the National Oceanic and Atmospheric Administration to establish a working group and issue a report on ocean farming practices and ways to deacidify oceans. (§761)
No comparable provision.	Directs USDA to submit a report to Congress describing how conservation programs may be used to conserve ocelots (wild cats native to the southwestern United States). (§762)

Table 4. (Continued)

House-reported H.R. 5961	Senate-passed H.R. 6147, Division C
No comparable provision.	Directs the Rural Housing Service to submit a report to Congress about the housing affected by Title V of the Housing Act of 1949, its loan status, and its affordability. (§763)
No comparable provision.	Directs the FDA to revise the advice in a notice about eating fish, to be consistent with nutrition science recognized by FDA about seafood consumption. (§764)
No comparable provision.	Amends implementation of a broadband program in the 2018 Agriculture Appropriation regarding eligibility determination and expansion efforts. (§765)

Source: CRS.

The explanatory statement that accompanies the final appropriation, and the House and Senate report language that accompanies the committee-reported bills, may also provide policy instructions. These documents do not have the force of law but often explain congressional intent, which the agencies are expected to follow.

Table 4 compares some of the major policy provisions that have been identified in the General Provisions (Title VII) of the FY2019 Agriculture appropriations bills. Many of these provisions have been included in past years' appropriations laws.

APPENDIX A. BUDGET SEQUESTRATION

Sequestration is a process to reduce federal spending through automatic, largely across-the-board reductions that permanently cancel mandatory and/or discretionary budget authority.[1] Sequestration is triggered as a budget enforcement mechanism when federal spending

[1] CRS Report R43411, The Budget Control Act of 2011: Legislative Changes to the Law and Their Budgetary Effects.

would exceed statutory budget goals.[2] Sequestration is currently authorized by the Budget Control Act of 2011 (BCA; P.L. 112-25).

Table A-1 shows the rates of sequestration that have been announced and the total amounts of budget authority that have been cancelled from accounts in Agriculture appropriations. *Table A-2* provides additional detail at the account level for mandatory accounts.

Table A-1. Sequestration from Accounts in Agriculture Appropriations (budget authority that was sequestered, in millions of dollars)

	Discretionary Accounts		Mandatory Accounts	
Fiscal Year	Rate	Amount	Rate	Amount
2013	5.0%	1,153	5.1%	713
2014	—	—	7.2%	1,052
2015	—	—	7.3%	1,153
2016	—	—	6.8%	1,819
2017	—	—	6.9%	1,686
2018	—	—	6.6%	1,316
2019	—	—	6.2%	1,530

Source: CRS, compiled from OMB, Reports to the Congress on the Joint Committee Reductions, various fiscal years. Available for FY2018-FY2019 at https://www. whitehouse.gov/ omb/legislative/sequestration-reports-orders and for FY2013-FY2017 at https://obamawhite house.archives.gov/omb/legislative_reports/sequest ration.

Notes: Sequestration rates listed here are for nonexempt, nondefense accounts. Amount totals were computed by CRS as compiled in Table A-2.

Discretionary Spending

For discretionary spending, sequestration is authorized through FY2021 if discretionary defense and nondefense spending exceed caps that are specified in statute (2 U.S.C. 901(c)).

[2] CRS Report R42972, Sequestration as a Budget Enforcement Process: Frequently Asked Questions.

In FY2013, the timing of the appropriations acts and the first year of sequestration resulted in triggering sequestration on discretionary spending.

In FY2014-FY2018, Bipartisan Budget Acts in 2013, 2015, and 2018 (BBAs; P.L. 113-67, P.L. 114-74, and P.L. 115-123, respectively) have avoided sequestration on discretionary spending. These BBAs raised the discretionary budget caps that were placed in statute by the BCA and allowed Congress to enact larger appropriations than would have been allowed.

For FY2019, the BBA in 2018 similarly provides a higher discretionary cap that may avoid sequestration (see "Discretionary Budget Caps and Subcommittee Allocations").

Mandatory Spending

Authorization of Sequestration

For mandatory spending, sequestration is presently authorized through FY2027, having been amended and extended by acts that were subsequent to the BCA (2 U.S.C. 901a(6)). That is, sequestration continues to apply annually to certain accounts of mandatory spending and is not avoided by the BBAs (*Table A-1*).

The original FY2021 sunset on the sequestration of *mandatory* accounts has been extended four times as an offset to pay for avoiding sequestration on *discretionary* spending in the near term or as a general budgetary offset for other authorization acts:

1. Congress extended the duration of mandatory sequestration by two years (until FY2023) as an offset in BBA 2013.[3]

[3] CBO, *Bipartisan Budget Act of 2013*, December 11, 2013, https://www.cbo.gov/publication/44964.

2. Congress extended it by another year (until FY2024) to maintain retirement benefits for certain military personnel (P.L. 113-82).
3. Congress extended sequestration on nonexempt mandatory accounts another year (until FY2025) as an offset in BBA 2015.[4]
4. Congress extended sequestration on nonexempt mandatory accounts by another two years (until FY2027) as an offset in BBA 2018 (P.L. 115-123 Division C, §30101(c)).[5]

Exemptions from Sequestration

Some farm bill mandatory programs are exempt from sequestration. Those expressly exempt by statute are the nutrition programs (SNAP, the child nutrition programs, and the Commodity Supplemental Food Program)[6] and the Conservation Reserve Program.[7] Some prior legal obligations in the Federal Crop Insurance Corporation[8] and the farm commodity programs may be exempt[9] as determined by OMB.[10]

Generally speaking, the experience since FY2013 is that OMB has ruled that most of crop insurance is exempt from sequestration, while the farm commodity programs, disaster assistance, and most conservation programs have been subject to it.[11]

---

[4] CBO, *Bipartisan Budget Act of 2015*, October 28, 2015, https://www.cbo.gov/publication/50938. In addition to extending sequestration and related to mandatory spending in agriculture, crop insurance was used as an additional budgetary offset in BBA 2015. The effect was temporary, however, and the crop insurance reduction was restored. For more background, see the section on crop insurance and the Standard Reinsurance Agreement in CRS Report R44240, *Agriculture and Related Agencies: FY2016 Appropriations*.

[5] CBO, *Bipartisan Budget Act of 2018*, February 8, 2018, https://www.cbo.gov/publication/53556.

[6] 2 U.S.C. 905 (h).

[7] 2 U.S.C. 905(g)(1)(A).

[8] 2 U.S.C. 905 (g)(2).

[9] 2 U.S.C. 906 (j).

[10] Some administrative expenses may be subject to sequestration, and therefore programs that are otherwise exempt may have a relatively small sequesterable amount compared to their total budget authority as shown, for example, in *Table A-2* for the Federal Crop Insurance Corporation, SNAP, child nutrition programs, and WIC.

[11] CRS Report R42050, Budget "Sequestration" and Selected Program Exemptions and Special Rules.

Implementation of Sequestration

Sequestration on nonexempt *mandatory* accounts continues in FY2019. Nonexempt mandatory spending is to be reduced by a 6.2% sequestration rate and thus paid at 93.8% of what would otherwise have been provided. This results in a reduction of about $1.5 billion from mandatory agriculture accounts in FY2019.

Table A-2. Sequestration of Mandatory Accounts in Agriculture Appropriations (sequestered budget authority in millions of dollars)

	FY 2013	FY 2014	FY 2015	FY 2016	FY 2017	FY 2018	FY 2019
Sequestration rate on nonexempt, nondefense mandatory accounts	5.1%	7.2%	7.3%	6.8%	6.9%	6.6%	6.2%
U.S. Department of Agriculture							
Office of the Secretary	—	—	0.9	0.9	0.9	0.9	0.8
Office of Chief Economist	—	—	0.1	0.1	0.1	0.1	0.1
Agricultural Research Service	0.1	0.1	0.1	0.1	0.1	0.1	0.1
National Institute of Food and Agriculture	—	—	—	9.9	10.0	—	—
Extension	0.3	0.4	1.8	—	—	3.3	0.3
Biomass Research and Development	—	—	0.2	0.2	0.2	—	—
Integrated Activities	—	—	7.3	—	—	6.6	5.0
Farm Production and Conservation Business Center	—	—	—	—	—	—	3.7
Animal and Plant Health Inspection Service							
Salaries appropriation	13.6	18.8	21.5	20.1	19.5	18.8	52.1
Miscellaneous Trust Funds	0.1	0.1	0.1	0.1	0.1	0.1	0.1
Food Safety Inspection Service	0.1	0.1	0.1	0.1	0.1	0.1	0.1
Grain Inspection Packers and Stockyards Admin.	2.1	3.0	3.0	3.1	3.1	2.9	—
Agricultural Marketing Service							
Section 32	40.4	79.7	81.9	77.3	79.6	78.1	74.4
Milk Market Orders Assessment Fund	2.9	4.2	4.2	4.0	4.1	4.0	3.8
Perishable Agriculture Commodities Act	0.6	0.8	0.8	0.8	0.8	0.7	0.7
Expenses and refunds	0.4	0.9	0.9	1.3	1.3	0.4	0.4

	FY 2013	FY 2014	FY 2015	FY 2016	FY 2017	FY 2018	FY 2019
Payments to states and possessions	—	—	5.3	5.0	5.0	5.6	5.3
Marketing services	—	—	2.2	2.0	2.1	2.0	3.0
Risk Management Agency	—	—	—	—	0.6	—	—
Federal Crop Insurance Corporation	3.0	4.2	5.9	3.5	3.9	3.7	5.0
Farm Service Agency							
Commodity Credit Corporation Fund	329.3	573.7	710.8	1,388.6	1,238.6	904.1	1,088.2
Agricultural Credit Insurance Corporation	—	—	0.1	0.1	0.1	0.1	0.1
Commodity Credit Corporation Export Loans	—	—	—	0.4	0.4	0.3	0.3
Pima Cotton Trust Fund	—	—	—	1.1	1.1	1.1	—
Wool Apparel Manufacturers Trust Fund	—	—	—	2.0	2.1	2.0	1.9
Agricultural Disaster Relief Fund	70.0	—	—	—	—	—	—
Tobacco Trust Fund	49.0	69.1	—	—	—	—	—
Natural Resources Conservation Service							
Farm bill conservation programs	171.2	263.1	269.9	265.7	281.5	254.9	262.4
Watershed Rehabilitation Program	—	11.9	11.2	4.7	4.7	4.4	3.0
Damage Assessment Restoration Revolving Fund	—	—	—	—	—	—	0.1
Rural Development, Rural Business Coop. Service							
Rural Energy for America Program	1.1	3.0	3.7	3.4	3.5	3.3	3.1
Economic Development Grants	—	—	—	—	—	—	0.9
Rural Microenterprise Investment Program	—	0.2	0.2	0.2	0.2	0.2	—
Energy Assistance Payments	3.3	3.2	1.1	1.0	1.0	1.0	—
Biorefinery Assistance Program	—	—	3.7	3.4	1.4	1.3	—
Rural Economic Development Grants	—	—	—	1.6	—	—	—
Foreign Agricultural Service	0.1	0.1	0.1	0.1	0.1	0.1	0.1

Table A-2. (Continued)

	FY 2013	FY 2014	FY 2015	FY 2016	FY 2017	FY 2018	FY 2019
			Food and Nutrition Service[a]				
SNAP	4.7	8.0	8.4	9.8	10.6	9.6	9.2
Child Nutrition Programs	2.5	4.2	4.2	3.9	4.3	4.0	3.9
Commodity Assistance Program	1.1	1.5	1.5	1.4	1.4	1.4	1.3
WIC	0.1	0.1	0.1	0.1	0.1	0.1	0.1
			Related Agencies				
			Food and Drug Administration				
Revolving Fund for Certification	0.4	0.6	0.6	0.6	0.6	0.6	0.6
User Fees	16.3	—	—	—	—	—	—
Farm Credit System Insurance Corporation	0.2	0.3	0.3	0.3	0.3	0.3	0.2
Commodity Futures Trading Commission	0.7	0.9	1.0	2.2	2.2	0.1	0.3
Total	713.3	1,051.9	1,153.0	1,818.9	1,685.6	1,315.9	1,530.3

Source: CRS, compiled from OMB, Reports to the Congress on the Joint Committee Reductions, various fiscal years. Available for FY2018-FY2019 at https://www.whitehouse.gov/omb/legislative/sequestration-reports-orders and for FY2013-FY2017 at https://obamawhitehouse.archives.gov/omb/legislative_reports/sequestration.

Notes: Sequestration rates are for nonexempt, nondefense accounts. The sequesterable budget authority for each cell may be computed by dividing the amount of sequestration by the sequestration rate. Column totals were computed by CRS based on the jurisdiction of the House Agriculture Appropriations Subcommittee.

[a] Benefits from the nutrition programs are generally exempt from sequestration by statute, but some administrative expenses in these programs may be subject to sequestration, and therefore a relatively small portion of the total budget authority may be sequesterable.

APPENDIX B. ACTION ON AGRICULTURE APPROPRIATIONS, FY1996-FY2019

Table B-1. Congressional Action on Agriculture Appropriations Since FY1996

	House Action			Senate Action			Final Appropriation			
Fiscal Year	Subcmte.	Cmte.	Floor	Subcmte.	Cmte.	Floor	Enacted[a]		Public Law	CRS Report
1996	6/14/1995	6/27/1995	7/21/1995	9/13/1995	9/14/1995	9/20/1995	10/21/1995	E	P.L. 104-37	95-624
1997	5/30/1996	6/6/1996	6/12/1996	7/10/1996	7/11/1996	7/24/1996	8/6/1996	E	P.L. 104-180	IB96015
1998	6/25/1997	7/14/1997	7/24/1997	7/15/1997	7/17/1997	7/24/1997	11/18/1997	E	P.L. 105-86	97-201
1999	6/10/1998	6/16/1998	6/24/1998	6/9/1998	6/11/1998	7/16/1998	10/21/1998	O	P.L. 105-277	98-201
2000	5/13/1999	5/24/1999	6/8/1999	6/15/1999	6/17/1999	8/4/1999	10/22/1999	E	P.L. 106-78	RL30201
2001	5/4/2000	5/16/2000	7/11/2000	5/4/2000	5/10/2000	7/20/2000	10/28/2000	E	P.L. 106-387	RL30501
2002	6/6/2001	6/27/2001	7/11/2001	Polled out[b]	7/18/2001	10/25/2001	11/28/2001	E	P.L. 107-76	RL31001
2003	6/26/2002	7/26/2002	—	7/23/2002	7/25/2002	—	2/20/2003	O	P.L. 108-7	RL31301
2004	6/17/2003	7/9/2003	7/14/2003	7/17/2003	11/6/2003	11/6/2003	1/23/2004	O	P.L. 108-199	RL31801
2005	6/14/2004	7/7/2004	7/13/2004	9/8/2004	9/14/2004	—	12/8/2004	O	P.L. 108-447	RL32301
2006	5/16/2005	6/2/2005	6/8/2005	6/21/2005	6/27/2005	9/22/2005	11/10/2005	E	P.L. 109-97	RL32904
2007	5/3/2006	5/9/2006	5/23/2006	6/20/2006	6/22/2006	—	2/15/2007	Y	P.L. 110-5	RL33412
2008	7/12/2007	7/19/2007	8/2/2007	7/17/2007	7/19/2007	—	12/26/2007	O	P.L. 110-161	RL34132
2009	6/19/2008	—	—	Polled out[b]	7/17/2008	—	3/11/2009	O	P.L. 111-8	R40000
2010	6/11/2009	6/18/2009	7/9/2009	Polled out[b]	7/7/2009	8/4/2009	10/21/2009	E	P.L. 111-80	R40721
2011	6/30/2010	—	—	Polled out[b]	7/15/2010	—	4/15/2011	Y	P.L. 112-10	R41475
2012	5/24/2011	5/31/2011	6/16/2011	Polled out[b]	9/7/2011	11/1/2011	11/18/2011	O	P.L. 112-55	R41964
2013	6/6/2012	6/19/2012	—	Polled out[b]	4/26/2012	—	3/26/2013	O	P.L. 113-6	R43110

	House Action			Senate Action			Final Appropriation			
Fiscal Year	Subcmte.	Cmte.	Floor	Subcmte.	Cmte.	Floor	Enacted[a]		Public Law	CRS Report
2014	6/5/2013	6/13/2013	—	6/18/2013	6/20/2013	—	1/17/2014	O	P.L. 113-76	R43110
2015	5/20/2014	5/29/2014	—	5/20/2014	5/22/2014	—	12/16/2014	O	P.L. 113-235	R43669
2016	6/18/2015	7/8/2015	—	7/14/2015	7/16/2015	—	12/18/2015	O	P.L. 114-113	R44240
2017	4/13/2016	4/19/2016	—	5/17/2016	5/19/2016	—	5/5/2017	O	P.L. 115-31	R44588
2018	6/28/2017	7/12/2017	9/14/2017	7/18/2017	7/20/2017	—	3/23/2018	O	P.L. 115-141	R45128
2019	5/9/2018	5/16/2018	—	5/22/2018	5/24/2018	8/1/2018	—	—	—	R45230

Source: CRS.

[a] E = Enacted as standalone appropriation (seven times over 22 years); O = Omnibus appropriation (13 times); Y = Year-long CR (two times).

[b] A procedure that permits a Senate subcommittee to transmit a bill to its full committee without a formal markup session. See CRS Report RS22952, Proxy Voting and Polling in Senate Committee.

In: Agricultural Issues
Editor: Michael Butlin

ISBN: 978-1-53615-474-0

Chapter 2

AGRICULTURAL CONSERVATION: A GUIDE TO PROGRAMS (UPDATED)*

Megan Stubbs

SUMMARY

The Natural Resources Conservation Service (NRCS) and the Farm Service Agency (FSA) in the U.S. Department of Agriculture (USDA) currently administer 20 programs and subprograms that are directly or indirectly available to assist producers and landowners who wish to practice conservation on agricultural lands. The differences and number of these programs have created general confusion about the purpose, participation, and policies of the programs. While recent consolidation efforts removed some duplication, a large number of programs remain. The programs discussed in this chapter are as follows

- Agricultural Conservation Easement Program (ACEP)
- Agricultural Management Assistance (AMA)
- Conservation Operations (CO); Conservation Technical Assistance (CTA)

* This is an edited, reformatted and augmented version of Congressional Research Service, Publication No. R40763, dated April 17, 2018.

- Conservation Reserve Program (CRP)
- CRP—Conservation Reserve Enhancement Program (CREP)
- CRP—Farmable Wetland Program
- CRP—Grasslands
- Conservation Stewardship Program (CSP)
- Emergency Conservation Program (ECP)
- Emergency Forest Restoration Program (EFRP)
- Emergency Watershed Protection (EWP)
- Environmental Quality Incentives Program (EQIP)
- EQIP—Conservation Innovation Grants (CIG)
- Grassroots Source Water Protection Program
- Healthy Forests Reserve Program (HFRP)
- Regional Conservation Partnership Program (RCPP)
- Voluntary Public Access and Habitat Incentive Program
- Water Bank Program
- Watershed and Flood Prevention Operations
- Watershed Rehabilitation Program

This tabular presentation provides basic information covering each of the programs. In each case, a brief program description is followed by information on major amendments in the Agricultural Act of 2014 (P.L. 113-79, 2014 farm bill), national scope and availability, states with the greatest participation, the backlog of applications or other measures of continuing interest, program funding authority, FY2018 funding, FY2019 Administration budget request, statutory authority, the authorization expiration date, and a link to the program's website.

INTRODUCTION

The U.S. Department of Agriculture (USDA) administers a number of agricultural conservation programs that assist private landowners with natural resource concerns. These include working land programs, land retirement and easement programs, watershed programs, emergency programs, technical assistance, and other programs. The number and funding levels for agricultural conservation programs have steadily increased over the past 60 years. Early conservation efforts undertaken by Congress were focused on reducing high levels of soil erosion and providing water to agriculture in quantities and quality that enhanced farm production. By the early 1980s, however, concern was growing that these programs were not adequately dealing with environmental problems resulting from agricultural activities (especially off the farm). In 1985, conservation policy took a new direction when Congress passed the Food Security Act of 1985 (1985 farm bill, P.L. 99-198), which established the first conservation programs designed to deal with environmental issues resulting from agricultural activities.

Provisions enacted in subsequent farm bills, including in 1990, 1996, 2002, 2008, and 2014,[1] reflect a rapid evolution of the conservation agenda, including the growing influence of environmentalists and other nonagricultural interests in the formulation of conservation policy, and a recognition that agriculture was not treated like other business sectors in many environmental laws.[2] Congress also began funding many of these new programs through mandatory spending for the first time, using the borrowing authority of USDA's Commodity Credit Corporation (CCC)[3] as the funding mechanism instead of annual appropriations. In addition to the original soil erosion and water quality and quantity issues, the conservation agenda has continued to expand to address other natural resource concerns, such as wildlife habitat, air quality, wetlands restoration and protection, energy efficiency, and sustainable agriculture.

Lead agricultural conservation agencies within USDA are the Natural Resources Conservation Service (NRCS), which provides technical assistance and administers most conservation programs, and the Farm Service Agency (FSA) which administers the Conservation Reserve Program (CRP). These agencies are supported by others in USDA that supply research and educational assistance, including the Agricultural Research Service (ARS), the Economic Research Service (ERS), the

---

[1] Conservation and Trade Act of 1990 (P.L. 101-624), Federal Agricultural Improvement and Reform Act of 1996 (P.L. 104-127), Farm Security and Rural Investment Act of 2002 (P.L. 107-171), Food, Conservation and Energy Act of 2008 (P.L. 110-246), and Agricultural Act of 2014 (P.L. 113-79).

[2] For additional discussion on conservation in the farm bill, see CRS Report R43504, *Conservation Provisions in the 2014 Farm Bill (P.L. 113-79)*.

[3] The CCC is the funding mechanism for the mandatory payments that are administered by various agencies of USDA, including all of the farm commodity price and income support programs. For more information on the CCC, see CRS Report R44606, *The Commodity Credit Corporation: In Brief*.

National Institute of Food and Agriculture (NIFA), and the Forest Service (FS).[4] In addition, agricultural conservation programs involve a large array of partners, including other federal agencies, state and local governments, and private organizations, among others, who provide funds, expertise, and other forms of assistance to further agricultural conservation efforts.

OVERVIEW

USDA provides technical and financial assistance to attract interest and encourage participation in conservation programs. Participation in all USDA conservation programs is voluntary. These programs protect soil, water, wildlife, and other natural resources on privately owned agricultural lands to limit environmental impacts of production activities both on and off the farm, while maintaining or improving production of food and fiber. Some of these programs center on improving or restoring resources that have been degraded, while others seek to create conditions with the objective of limiting degradation in the future.

Though programs in this chapter are listed alphabetically, agricultural conservation programs can be grouped into the following categories based on similarities: working land programs, land retirement and easement programs, watershed programs, emergency programs,[5] compliance,[6] technical assistance, and other programs and overarching provisions.

---

[4] For more information on ARS projects, see http://www.ars.usda.gov/Research/Research.htm; ERS projects, see http://www.ers.usda.gov/topics/natural-resources-environment/conservation-programs.aspx; NIFA programs, see http://www.nifa.usda.gov/nea/nre/nre.cfm; and FS projects, see http://www.fs.fed.us/projects/.

[5] Additional information on emergency land rehabilitation programs may be found in CRS Report R42854, *Emergency Assistance for Agricultural Land Rehabilitation*.

[6] Compliance refers to a series of farm bill provisions that require a minimum level of conservation on environmentally sensitive land in exchange for access to other USDA program benefits. Compliance provisions are not discussed in this report. Additional analysis may be found in CRS Report R42459, *Conservation Compliance and U.S. Farm Policy*.

USDA Agricultural Conservation Programs

Working Lands Programs—typically classified as programs that allow private land to remain in production, while implementing various conservation practices to address natural resource concerns specific to the area.

- Environmental Quality Incentives Program (EQIP), Conservation Stewardship Program (CSP), and Agricultural Management Assistance (AMA).

Land Retirement and Easement Programs—land retirement programs provide federal payments to private agricultural landowners for *temporary* changes in land use or management to achieve environmental benefits. Conversely, conservation easements impose a *permanent* or *long-term* land-use restriction that is voluntarily placed on the land in exchange for a government payment.

- Conservation Reserve Program (includes the Conservation Reserve Enhancement Program [CREP], Farmable Wetland Program, and Grasslands), Agricultural Conservation Easement Program (including agricultural land easements and wetland reserve easements), and Healthy Forests Reserve Program (HFRP).

Watershed Programs—NRCS partners with local sponsors to carry out activities for soil conservation; flood prevention; conservation, development, utilization, and disposal of water; watershed surveys; and dam rehabilitation.

- Watershed and Flood Prevention Operations (also referred to as the Small Watershed Program or P.L. 566 and P.L. 534) and Watershed Rehabilitation program.

Emergency Programs—provide disaster assistance for farmland rehabilitation and impairments to watersheds. Programs are usually funded through supplemental appropriation acts.

- Emergency Conservation Program (ECP), Emergency Forest Restoration Program (EFRP), and Emergency Watershed Protection (EWP) program (includes floodplain easements).

Compliance—provisions that prohibit a producer from receiving many federal farm program benefits (including conservation assistance and crop insurance) when conservation program requirements for highly erodible lands and wetlands are not met.

- Highly erodible land conservation (Sodbuster), wetland conservation (Swampbuster), and Sodsaver.

Technical Assistance Programs—provides landowners with science-based conservation information and technical expertise (e.g., engineering and biological) unique to the region and land use type. Usually does not include financial assistance.

- Conservation Operations (includes Conservation Technical Assistance, Soil Survey, Snow Survey and Water Supply Forecasting, and Plant Materials Centers).

Other Conservation Programs and Provisions—Conservation Innovation Grants, Grassroots Source Water Protection Program, Regional Conservation Partnership Program (RCPP), Voluntary Public Access and Habitat Incentive Program, and Water Bank Program.

The majority of conservation programs are funded through USDA's Commodity Credit Corporation (CCC) as mandatory spending. Congress authorizes mandatory programs at specified funding levels (or acreage enrollment levels for CRP and CSP) for multiple years, typically through omnibus legislation such as the farm bill. Mandatory programs are funded at these levels unless Congress limits funding to a lower amount through the appropriations or legislative process (or puts a ceiling on acreage that can be enrolled).[7] Discretionary programs are funded each year through the annual appropriations process.[8]

Despite a steady increase in mandatory funding authority, select conservation programs have been reduced or capped through annual appropriation acts since FY2003. Many of these spending reductions were at the request of the Administration. The mix of programs and amount of reductions vary from year to year. Some programs, such as the CRP, have

---

[7] For additional discussion on these reductions, see CRS In Focus IF10041, *Reductions to Mandatory Agricultural Conservation Programs in Appropriations Law*.

[8] For additional information on the FY2018 appropriation, see CRS Report R45128, *Agriculture and Related Agencies: FY2018 Appropriations*.

not been reduced by appropriators in recent years, while others, such as EQIP, have been repeatedly reduced below authorized levels. Authorized mandatory funding for conservation programs has been reduced by a total of more than $4 billion over the past 10 years. FY2018 marks the first time in 15 years that an appropriations act does not reduce mandatory conservation program funding.

Sequestration has also had an effect on conservation programs. Sequestration is a process of automatic, largely across-the-board reductions that permanently cancel mandatory and/or discretionary budget authority to enforce statutory budget goals.[9] Discretionary accounts have avoided sequestration in recent years through adjustments to spending limits, although sequestration continues on mandatory accounts. Most all mandatory conservation programs were subject to sequestration in FY2014 through FY2018.[10] Even with sequestration and appropriations act reductions, total annual mandatory funding for conservation programs has grown from a total of $3.5 billion in FY2008 to over $5 billion in FY2018.

2014 Farm Bill

Before the 1985 farm bill, few conservation programs existed, and only two would be considered large by today's standards. In contrast, leading up to the debate on the 2014 farm bill, there were over 20 distinct conservation programs with total annual spending greater than $5 billion. The differences and number of these programs created general confusion about the purpose, participation, and policies of the programs. Discussion

---

[9] For additional information on sequestration and the budget process, see CRS Report R42972, *Sequestration as a Budget Enforcement Process: Frequently Asked Questions*.

[10] The Conservation Reserve Program is statutorily exempt from sequestration (2 U.S.C. 905 (g)(1)(A)). Sequestration reductions in FY2014 were reflected in the Congressional Budget Office (CBO) baseline that was used to write the 2014 farm bill. For more information, see CRS Report R42484, *Budget Issues That Shaped the 2014 Farm Bill*. Sequestration estimates may be found in the Office of Management and Budget's (OMB) Reports to the Congress on the Joint Committee Reductions can be found at https://www.whitehouse.gov/omb/legislative/sequestration-reportsorders/. Sequestration will remain in effect through FY2027 unless there is congressional action.

about simplifying or consolidating conservation programs to reduce overlap and duplication, and to generate savings, continued for a number of years. The Agricultural Act of 2014 (P.L. 113-79, 2014 farm bill), contained several program consolidation measures, including the repeal of 12 active and inactive programs, the creation of two new programs, and the merging of two programs into existing ones.[11]

UNFUNDED AND REPEALED PROGRAMS

A number of conservation programs were repealed by the 2014 farm bill or have gone unfunded by Congress in recent years. Table 1 lists these programs and the most recent congressional action taken.

Table 1. Unfunded and Repealed Programs

Program Name	Action-Year	Description
Agricultural Water Enhancement Program (AWEP)	Repealed-2014	A subprogram of EQIP that provided funding for water quality and quantity projects. Repealed in the 2014 farm bill and incorporated into RCPP.
Chesapeake Bay Watershed Program	Repealed-2014	Provided additional funding through existing programs to conservation projects in the Chesapeake Bay watershed. Repealed in the 2014 farm bill and incorporated into RCPP.
Conservation Security Program	New enrollment unauthorized-2008	Replaced by the Conservation Stewardship Program in the 2008 farm bill, the program enrolled acres in 5- to 10-year stewardship contracts, the last of which will expire in FY2018.
Farmland Protection Program (FPP)	Repealed-2014	An agricultural land easement program repealed in the 2014 farm bill. Program components were incorporated into the Agricultural Conservation Easement Program (ACEP).
Grasslands Reserve Program (GRP)	Repealed-2014	A grassland easement and contract program repealed in the 2014 farm bill. Easement provisions were incorporated into ACEP and rental contracts were incorporated into CRP.

[11] For additional information on the amendments to conservation programs in the 2014 farm bill, see CRS Report R43504, *Conservation Provisions in the 2014 Farm Bill (P.L. 113-79)*.

Program Name	Action-Year	Description
Resource Conservation and Program (RC&D) Development	Unfunded since-FY2011	Funded local RC&D coordinator positions. Funding terminated in FY2011 and program close-out complete in FY2012. FY2014 appropriations act permanently cancelled any remaining funds.
Watershed Surveys and Planning	Unfunded since-FY2006	Funded investigations and surveys of river basins to respond to water quality, flooding, water and land management, and sedimentation problems.
Wetland Reserve Program (WRP)	Repealed-2014	A wetland easement program repealed in the 2014 farm bill. Program components were incorporated into ACEP.
Wildlife Habitat Incentives Program	Repealed-2014	A wildlife habitat cost-share assistance program repealed in the 2014 farm bill. Program components were incorporated into EQIP.

Source: CRS.

CONSERVATION PROGRAMS

The tabular presentation that follows provides basic information covering each of the USDA agricultural conservation programs, including

- administering agency or agencies within USDA;
- brief program description;
- major amendments to the program in the Agricultural Act of 2014 (P.L. 113-79), commonly referred to as the 2014 farm bill;
- national scope and availability, including participation levels and acres enrolled;
- states with the highest level of funds obligated or acres enrolled;
- volume of application backlog or public interest in each program;
- authorized funding levels, whether mandatory spending or discretionary appropriations, and any funding restrictions;
- FY2018 funding level in the Consolidated Appropriations Act of 2018 (P.L. 115-124), or, if applicable, the authorized level in the Agricultural Act of 2014 (sequestration and carryover not included unless noted);
- FY2019 funding level requested by the Administration (sequestration and carryover included where known);

- statutory authority, recent amendments, and U.S. Code reference;
- expiration date of program authority unless permanently authorized; and
- program's website link.

Information for the following tables is drawn from agency budget presentations, explanatory notes, and websites; written responses to questions published each year in hearing records of the Agriculture Appropriations Subcommittees of the House and Senate Appropriations Committees; and spending estimates from the Congressional Budget Office.

Table 2. Agricultural Conservation Easement Program (ACEP)

Administering agency(s)	**NRCS**
Program description	ACEP provides financial and technical assistance through two types of easements: agricultural land easements that limit nonagricultural uses on productive farm or grass lands, and wetland reserve easements that protect and restore wetlands. *Agricultural Land Easements (ALE)*—Enrollment is through eligible entities that enter into cooperative agreement to obligate ACEP funds. The entities acquire easements and hold, monitor, manage, and enforce the easements. The federal share of easement acquisition cannot exceed 50% of the appraised fair market value or 75% if it is determined to be a grassland of special environmental significance. *Wetland Reserve Easement (WRE)*—Enrollment options (federal share) include permanent easements (100% easement value and 75%-100% restoration cost), 30-year easements (50%-75% easement value and 50%-75% restoration cost), term easements or the maximum duration under state law (50%-75% easement value and 50%-75% restoration cost), and 30-year contracts only available to Indian tribes (50%-75% easement value and 50%-75% restoration cost). NRCS pays all costs associated with recording the easement.
Major 2014 farm bill amendments	Created in the 2014 farm bill from three repealed programs—Farmland Protection Program, Grassland Reserve Program, and Wetlands Reserve Program. General program provisions are the same across both easement types, including ineligible land; subordination, exchange, modification, and termination procedures; and compliance requirements.

Administering agency(s)	**NRCS**
National scope	Available nationwide. In FY2017, $214.1 million in ACEP funding was used to enroll an estimated total of nearly 300,000 acres of farmland, grassland, and wetlands through 659 new ACEP easements. This includes ALE (easements)––186 on 79,487 acres; ALE (grassland easements)—27 on 127,148 acres; WRE (permanent easements)—397 on 72,386 acres; and WRE (30-year easements)—47 on 19,479 acres.
Leading states	The highest levels of funding obligated in FY2017 were in FL ($51.4 million), AR ($46.4 million), and MT ($31.7 million).
Backlog/ Interest	*Agricultural Land Easements (ALE)*—In FY2017, 542 ALE applications were received for over 336,800 acres, including 39 applications for over 168,600 acres of grasslands of special environmental significance. Approximately 39% of applications were enrolled. *Wetland Reserve Easements (WRE)*—In FY2017, 2,336 WRE applications were for over 352,870 acres. Approximately 19% of applications were funded.
Funding authority	Mandatory, subject to sequestration. FY2014—$400 million, FY2015—$425 million. FY2016—$450 million, FY2017—$500 million, and FY2018—$250 million.
FY2018 funding	$233.5 million (authorization reduced by approximately $16.5 million from sequestration).
FY2019 Administration request	$234.5 million (authorization reduced by approximately $15.5 million from sequestration), subject to reauthorization.
Statutory authority	Authorized in subtitle D of Title II (§2301) of the Agricultural Act of 2014 (P.L. 113-79) as §1265 of the Food Security Act of 1985 (P.L. 99-198), as amended. 16 U.S.C. §§3865-3865d.
Authorization expires	September 30, 2018.
Program website	http://www.nrcs.usda.gov/wps/portal/nrcs/main/national/programs/easements/acep/

Further information about these programs may be found on the NRCS website at http://www.nrcs.usda.gov and on the "conservation programs" page of the FSA website at http://www.fsa.usda.gov.

Table 3. Agricultural Management Assistance (AMA)

Administering agency(s)	**NRCS (conservation assistance), Agricultural Marketing Service (AMS, organic certification), and Risk Management Agency (RMA, production, price, or revenue risk reduction).**
Program description	AMA provides cost-sharing assistance under contracts of one to 10 years to producers in 16 specified states where participation in the federal crop insurance program has been historically low. Producers use this assistance to construct or improve water management and irrigation structures, plant trees, control soil erosion, practice integrated pest management, practice organic farming, develop value-added processing, and enter into futures, hedging, or options contracts to reduce production, price, or revenue risk.
Major 2014 farm bill amendments	Both the House- and Senate-passed farm bills (H.R. 2642 and S. 954) included amendments to AMA, but none were adopted in the conference agreement.
National scope	Not available nationwide. Eligible states include CT, DE, HI, ME, MD, MA, NV, NH, NJ, NY, PA, RI, UT, VT, WV, and WY. In FY2017, over $4.7 million in AMA funding was used to enroll 377 AMA contracts covering over 2,000 acres within eligible states.
Leading states	States with the most funds obligated (for conservation only) in FY2017 include ME ($572,000), PA ($494,000), and MD ($248,000).
Backlog/ Interest	In FY2017, AMA had a backlog of 644 pending applications that would enroll more than 4,200 acres. The highest numbers of unfunded applications were submitted in ME (234), WV (161), and PA (94).
Funding authority	Mandatory, subject to sequestration. Permanently authorized at $10 million for each fiscal year. Funding is split by law among the three USDA agencies: 50% to NRCS, 10% to AMS, and 40% to RMA.
FY2018 funding	$9.3 million ($4.7 million for conservation; total authorization reduced by approximately $660,000 from sequestration).
FY2019 Administration request	$0. Requests a rescission of the reduced total for FY2018 ($10 million minus sequestration).
Statutory authority	Authorized in Title I, §133 of the Agricultural Risk Protection Act of 2000 (P.L. 106-224) as §524(b) of the Federal Crop Insurance Act, as amended. 7 U.S.C. §1524(b).
Authorization expires	Permanent authorization.
Program websites	http://www.nrcs.usda.gov/wps/portal/nrcs/main/national/programs/financial/ama (NRCS); http://www.ams.usda.gov/services/grants/occsp (AMS).

Table 4. Conservation Operations (CO) — Conservation Technical Assistance (CTA)

Administering agency(s)	**NRCS**
Program description	Conservation Operations (CO) is the primary account funding technical assistance within NRCS. More than 88% of CO funding is for Conservation Technical Assistance (CTA), which provides conservation planning and implementation assistance through field staff placed in almost all counties within the United States and territories. This assistance is provided to producers and landowners who voluntarily apply natural resource conservation systems, consisting of one or more practices, on private and other nonfederal lands. Other components of CO include the Soil Surveys, Snow Survey and Water Supply Forecasting, and Plant Materials Centers.
Major 2014 farm bill amendments	None.
National scope	Available nationwide. CTA was funded at $759 million and 4,347 staff years for FY2017. Total CO spending for FY2017 was $864 million and 4,849 staff years.
Leading states	No data are available for the CTA subset in FY2017, but the three leading states for total CO funding are TX ($31.3 million), MO ($25.2 million), and IA ($22.2 million).
Backlog/ Interest	Not available.
Funding authority	Discretionary. No specific authorization level.
FY2018 funding	$774.4 million for CTA out of $874.1 million appropriated for all CO.
FY2019 Administration request	$575.9 million for CTA out of $669.0 million for all CO.
Statutory authority	Authorized in the Soil Conservation and Domestic Allotment Act (P.L. 74-46), as amended. 16 U.S.C. §590a-g, 16 U.S.C. §590q.
Authorization expires	Permanent authorization.
Program website	http://www.nrcs.usda.gov/programs/cta/ (CTA); http://www.nrcs.usda.gov/wps/portal/nrcs/site/soils/home/ (soil survey); http://www.wcc.nrcs.usda.gov/ (snow survey and water supply forecasting); and http://plant-materials.nrcs.usda.gov/ (plant materials centers).

Table 5. Conservation Reserve Program (CRP)

Administering agency(s)	**FSA, with technical assistance by NRCS.**
Program description	CRP provides annual rental payments, usually over 10 years, to producers to replace crops on highly erodible and environmentally sensitive land with long-term resource-conserving plantings. Bids to enroll land are solicited during a limited time period, then compared using an Environmental Benefits Index (EBI). Those with the highest EBI scores are accepted. This is referred to as general sign-up. Embedded in the CRP are several small and more focused subprograms and initiatives—some established in law and others established administratively—that bypass the general bidding process and address specific resource topics. Examples of these focus on concentrated resource problems in a portion of a state, protection of small isolated agricultural wetlands, or improvement of habitat for upland game birds. All lands that qualify for these subprograms and initiatives are automatically accepted and enrolled on a continuous basis. This is referred to as continuous sign-up.
Major 2014 farm bill amendments	Reduces enrollment ceiling from 32 million acres to 24 million acres by FY2018; amends emergency harvesting, grazing, and permits other use of forage, in some cases, without a reduction in rental rate; allows a one-time, penalty-free, early out in FY2015 for contracts enrolled longer than five years and containing no environmentally sensitive practices; and incorporates grassland contracts, similar to what was repealed under the Grassland Reserve Program (GRP). These grassland contracts are referred to as grassland sign-up.
National scope	At the end of FY2017, there were 638,723 active contracts on 358,673 farms with 23.4 million acres enrolled. Of this total, 16.0 million acres are enrolled under general sign-up, 1.1 million acres under Conservation Reserve Enhancement Program sign-up (CREP), 5.9 million acres under non-CREP sign-up, 393,796 acres under Farmable Wetland sign-up, and 89,052 acres under grasslands sign-up.
Leading states	Leading states by total acres enrolled are TX (2.9 million), KS (2.1 million), and IA (1.8 million). Leading states by total number of contracts are IA (109,068), IL (78,903), and MN (55,026).
Backlog/ Interest	Interest in the most recent general sign-up (#49) was high, with a 22% acreage offer acceptance rate (1.9 million acres offered and 407,416 acres accepted). No general sign-up is expected in FY2018. An estimated 1.5 million acres are expected to expire from the program at the end of FY2018 (including 440,000 continuous and 1.1 million general sign-up).
Funding authority	Mandatory, statutorily exempt from sequestration. At any one time, CRP can enroll no more than: 27.5 million acres in FY2014; 26 million acres in FY2015; 25 million acres in FY2016; and 24 million acres in FY2017 and FY2018. No funding amount specified.
FY2018 est. funding	$1.8 billion (based on the estimated number of acres that will be enrolled, including technical assistance).

Administering agency(s)	**FSA, with technical assistance by NRCS.**
FY2019 est. Administration request	$1.9 billion (based on the estimated number of acres that will be enrolled, including technical assistance). Request also proposes refocusing the program, eliminating select incentive payments, and limiting annual rental payments to 80% of National Agricultural Statistical Service county rental rates.
Statutory authority	Authorized in §§1231-1235 of the Food Security Act of 1985 (P.L. 99-198), as amended. Amended in §§2001-2008 of the Agricultural Act of 2014 (P.L. 113-79). 16 U.S.C. §§3831(a)- 3835a.
Authorization expires	September 30, 2018.
Program website	http://www.fsa.usda.gov/programs-and-services/conservation-programs/conservation-reserve- program/

Table 6. CRP—Conservation Reserve Enhancement Program (CREP)

Administering agency(s)	**FSA, with technical assistance by NRCS.**
Program description	This subprogram of CRP partners with states at their request. States propose substate areas such as a watershed, where environmental or resource concerns are more concentrated and can be addressed by enrolling up to 100,000 acres per project. States contribute 20% of the funding to allow for larger payments, in order to encourage greater participation. Average rental payments are higher than for acreage under the general CRP sign-up process. Sign-up is held on a continuous basis.
Major 2014 farm bill amendments	None.
National scope	There are 47 CREP agreements in 34 states, including 64,057 contracts on 43,071 farms, enrolling a total of 1.1 million acres, as of September 2017.
Leading states	Leading states by number of acres enrolled are PA (142,777), OH (120,051), and KY (97,456). States leading in number of contracts are OH (15,413), PA (8,769), and MD (4,734).
Backlog/ Interest	In May 2017, FSA announced a freeze on all continuous sign-up due to the 24 million acreage enrollment limit. Between May and September 2017, 1.185 million acres of continuous sign-up (including CREP acres) were approved for acceptance but not enrolled because of the freeze.
Funding authority	Unspecified acreage subset of CRP.
FY2017 funding	Unspecified acreage subset of CRP.
FY2018 Administration request	Unspecified acreage subset of CRP.
Statutory authority	Authority derived from CRP statutory authority (see "Conservation Reserve Program (CRP)").
Authorization expires	September 30, 2018.
Program website	http://www.fsa.usda.gov/programs-and-services/conservation-programs/conservation- reserve-enhancement/index

Table 7. CRP—Farmable Wetland Program

Administering agency(s)	**FSA, with technical assistance by NRCS.**
Program description	This 750,000 acre subprogram of the CRP enrolls small isolated agricultural wetlands. On a single tract of land, enrollment is set at a maximum of 40 contiguous wetland acres. "Flooded farmland" has a 20-acre limit. Eligible lands include wetlands that were cropped in three of the preceding 10 years (and include buffers sufficient to protect them), on which the hydrology will be restored and a vegetative cover established. Sign-up is held on a continuous basis.
Major 2014 farm bill amendments	Renames the pilot program "Farmable Wetland Program." Reauthorizes the program through FY2018, and clarifies language related to constructed wetlands receiving water from agricultural drainage. Reduces acreage limitation from 1 million acres to 750,000 acres.
National scope	Active contracts in 22 participating states, including AL, AR, CO, ID, IL, IN, IA, KS, LA, MI, MN, MS, MO, MT, NE, NC, ND, OH, OK, SD, WA, and WI. As of September 2017, there are 15,785 contracts on 12,087 farms for a total of 393,796 enrolled acres.
Leading states	Leading states by acres enrolled are ND (130,430), SD (100,587), and IA (89,517). The largest number of contracts are in IA (5,065), followed by SD (4,163) and ND (2,889).
Backlog/ Interest	In May 2017, FSA announced a freeze on all continuous sign-up due to the 24 million acreage enrollment limit. Between May and September 2017, 1.185 million acres of continuous sign-up (including the Farmable Wetland Program acres) were approved for acceptance but not enrolled because of the freeze.
Funding authority	Mandatory, statutorily exempt from sequestration. No more than 750,000 acres enrolled at any one time and no more than 100,000 acres in any state (may be increased to 200,000 acres after agency review).
FY2017 funding	Unspecified acreage subset of CRP.
FY2018 Administration request	Unspecified acreage subset of CRP.
Statutory authority	Authorized in Title XI of Agriculture and Related Agency appropriations, 2001 (P.L. 106- 387) as §1231B of the Food Security Act of 1985 (P.L. 99-198), as amended. Amended by §2002 of the Agricultural Act of 2014 (P.L. 113-79). 16 U.S.C. §3831b.
Authorization expires	September 30, 2018.
Program website	http://www.fsa.usda.gov/programs-and-services/conservation-programs/farmable-wetlands/ index

Table 8. CRP—Grasslands

Administering agency(s)	**FSA, with technical assistance by NRCS.**
Program description	This subprogram of the CRP enrolls grassland—including rangeland, pastureland, and certain other lands—while maintaining the area as grazing. Contracts are for 14 and 15 years. Rental payments are limited to 75% of the contracted land's grazing value. Small livestock operations are considered in a separate ranking pool during enrollment. Expiring CRP acres have priority enrollment. Sign-up is held on a continuous basis.
Major 2014 farm bill amendments	Created in the 2014 farm bill, partially from the repealed Grassland Reserve Program.
National scope	Active contracts in 20 states, including CO, IN, IA, KS, KY, MN, MO, MT, NE, NM, NC, ND, OH, OK, PA, SD, TN, TX, WI, and WY. As of September 2017, there are 309 contracts on 196 farms for a total of 89,052 enrolled acres.
Leading states	Leading states by acres enrolled are NE (25,115), SD (14,979), and CO (13,214). The largest number of contracts are in NE (91), followed by SD (56) and ND (28).
Backlog/ Interest	The CRP grasslands sign-up (#200) had an acreage acceptance rate of 10% (1.0 million acres offered and 101,293 acres accepted). Applicants in all states except seven (AK, DE, ME, NV, NH, RI, and VT) offered acres. Approximately 800,000 acres of grassland sign-up had been approved for contracts beginning in FY2018, with only 420,000 acres completely enrolled as of September 2017. In May 2017, FSA announced a freeze on all continuous sign-up due to the 24 million acreage enrollment limit. Between May and September 2017, 1.185 million acres of continuous sign-up (including CRP Grassland acres) were approved for acceptance but not enrolled because of the freeze.
Funding authority	Mandatory, statutorily exempt from sequestration. No more than 2 million acres enrolled at any one time between FY2014 and FY2018.
FY2017 funding	Unspecified acreage subset of CRP.
FY2018 Administration request	Unspecified acreage subset of CRP.
Statutory authority	Authorized in subtitle A of Title II (§2001 *et seq.*) of the Agricultural Act of 2014 (P.L. 113- 79) as §1231 *et seq.* of the Food Security Act of 1985 (P.L. 99-198), as amended. 16 U.S.C. §3831 *et seq.*
Authorization expires	September 30, 2018.
Program website	https://www.fsa.usda.gov/programs-and-services/conservation-programs/crp-grasslands/index

Table 9. Conservation Stewardship Program (CSP)

Administering agency(s)	NRCS
Program description	CSP provides financial and technical assistance to promote the conservation and improvement of soil, water, air, energy, plant and animal life, and other conservation purposes on tribal and private working lands. Contracts (five years in length with the option of renewal) are based on meeting or exceeding a "stewardship threshold." CSP provides two possible payments: (1) an annual payment for installing new conservation activities and maintaining existing activities and (2) a supplemental payment for adopting a resource-conserving crop rotation. Enrollment is offered through a continuous sign-up and applications are accepted year-round.
Major 2014 farm bill amendments	Reduces the enrollment cap from 12.769 million acres annually to 10 million acres annually. Reorganizes the statutory language and refocuses the program on generating additional conservation benefits. Raises the entry bar for participants to two priority resource concerns upon entry and requires the participant to meet or exceed one additional priority resource concern by the end of the contract. Contract renewal participants must meet the threshold for two additional priority resources concerns, or exceed the threshold for two existing priority resource concerns. Removes the 10% limitation on nonindustrial private forest land and provides flexible transition options for land coming out of CRP.
National scope	Available nationwide. The program held its first sign-up in 2009, and at the end of FY2017, more than 87.3 million acres were enrolled. Of the expiring FY2012 contracts, 57% were renewed in FY2017 for an additional five years.
Leading states	In FY2017, SD had the most total acres funded (661,494), followed by NE (637,159) and ND (534,333). The most funding obligated in FY2017 was in MS ($7.3 million), SD ($7.1 million), and NE ($5.1 million).
Backlog/ Interest	In FY2017, CSP had a backlog of 4,620 valid applications that would enroll approximately 4.4 million acres. The highest numbers of unfunded applications were submitted in AR (480), OK (403), and MS (375). The highest numbers of unfunded applications by acres were in MT (591,622), AR (448,388), and ND (433,654).
Funding authority	Mandatory, subject to sequestration. CSP can enroll up to 10 million acres each fiscal year. No funding amount specified.
FY2018 est. funding	$1.3 billion; 10 million acres (authorization reduced by sequestration; unknown effect on acres).
FY2019 est. Administration request	Proposes to eliminate new enrollments for FY2019, subject to reauthorization.
Statutory authority	Authorized in §2301 of the Food, Conservation, and Energy Act of 2008 (P.L. 110-246) as §§1238D-1238G of the Food Security Act of 1985 (P.L. 99-198), as amended. Amended in §2101 of the Agricultural Act of 2014 (P.L. 113-79). 16 U.S.C. §§3838d-3838g.
Authorization expires	September 30, 2018.
Program website	http://www.nrcs.usda.gov/wps/portal/nrcs/main/national/programs/financial/csp

Table 10. Emergency Conservation Program (ECP)

Administering agency(s)	**FSA, with technical assistance by NRCS.**
Program description	ECP provides emergency funding and technical assistance to producers to rehabilitate farmland damaged by natural disasters (e.g., hurricanes, floods, wind, and erosion) through activities such as removing debris, and implementing emergency water conservation measures in response to severe droughts.
Major 2014 farm bill amendments	None.
National scope	Available nationwide. Participation and funding varies widely from year to year.
Leading states	Not applicable.
Backlog/ Interest	Not applicable.
Funding authority	Discretionary. No specific authorization level.
FY2018 funding to date	The Bipartisan Budget Act of FY2018 (P.L. 115-123) provided $400 million to remain available until expended for necessary expenses related to Hurricanes Harvey, Irma, and Maria and wildfires occurring in calendar year 2017.
FY2019 Administration request	$0
Statutory authority	Authorized in §401 of the Agriculture Credit Act of 1978 (P.L. 95-334), as amended. 16 U.S.C. §§2201-2205.
Authorization expires	Permanent authorization.
Program website	http://www.fsa.usda.gov/programs-and-services/conservation-programs/emergency- conservation/index

Table 11. Emergency Forest Restoration Program (EFRP)

Administering agency(s)	**FSA, with technical assistance by NRCS.**
Program description	EFRP provides cost-share assistance to private forestland owners to repair and rehabilitate damage caused by a natural disaster on nonindustrial private forest land. Natural disasters include wildfires, hurricanes or excessive winds, drought, ice storms or blizzards, floods, or other resource-impacting events, as determined by USDA.
Major 2014 farm bill amendments	None.
National scope	Available nationwide. Participation and funding varies widely from year to year.
Leading states	Not applicable.
Backlog/Interest	Not applicable.
Funding authority	Discretionary. No specific authorization level.
FY2018 funding to date	$0

Table 11. (Continued)

Administering agency(s)	**FSA, with technical assistance by NRCS.**
FY2019 Administration request	$0
Statutory authority	Authorized in §8203 of the Food, Conservation, and Energy Act of 2008 (P.L. 110-246) as §407 of the Agriculture Credit Act of 1978 (P.L. 95-334). 16 U.S.C. §2206.
Authorization expires	Permanent authorization.
Program website	http://www.fsa.usda.gov/programs-and-services/disaster-assistance-program/emergency- forest-restoration/index

Table 12. Emergency Watershed Protection (EWP)

Administering agency(s)	**NRCS on private lands and U.S. Forest Service on National Forest Systems lands.**
Program description	EWP provides technical and financial assistance to reduce hazards to life and property in watersheds that have been damaged by natural disasters. Assistance includes disaster cleanup and recovery activities, and purchasing easements in floodplains that will benefit natural resources such as wetlands, while reducing the risk of exposure to future natural disasters.
Major 2014 farm bill amendments	Authorizes USDA to modify and terminate floodplain easements provided the current landowner agrees, and the modification or termination addresses a compelling public need for which there is no practical alternative, and is in the public interest.
National scope	Available nationwide. Participation and funding varies widely from year to year.
Leading states	Not applicable.
Backlog/ Interest	Not applicable.
Funding authority	Discretionary. No specific authorization level.
FY2018 funding to date	The Bipartisan Budget Act of FY2018 (P.L. 115-123) provided $541 million to remain available until expended for necessary expenses related to Hurricanes Harvey, Irma, and Maria and wildfires occurring in calendar year 2017.
FY2019 Administration request	$0
Statutory authority	Authorized in §216 of P.L. 81-516 and §403 of the Agriculture Credit Act of 1978 (P.L. 95- 334), as amended. Amended in §2506 of the Agricultural Act of 2014 (P.L. 113-79). 16 U.S.C. §2203; and 33 U.S.C. §701b-1.
Authorization expires	Permanent authorization.
Program website	http://www.nrcs.usda.gov/wps/portal/nrcs/main/national/programs/landscape/ewpp

Table 13. Environmental Quality Incentives Program (EQIP)

Administering agency(s)	**NRCS**
Program description	EQIP provides financial and technical assistance to producers and land owners to plan and install structural, vegetative, and land management practices on eligible lands to alleviate natural resource problems. Eligible producers enter into contracts to receive payment for implementing conservation practices. Approved activities are carried out according to an EQIP plan developed in conjunction with the producer that identifies the appropriate conservation practice(s) to address specific land resource concerns. Sixty percent of the funds are targeted to conservation practices benefiting livestock.
Major 2014 farm bill amendments	Incorporates the Wildlife Habitat Incentives Program (WHIP) into EQIP with a 5% allocation to wildlife habitat practices; removes the minimum one-year contract length requirement; adds veteran farmer or rancher to the list of certain producers eligible for cost-share rates up to 90% and advanced payments; raises the payment limit to an aggregate of $450,000 between FY2014-FY2018 and eliminates the waiver authority for contracts of environmental significance; repeals the Agricultural Water Enhancement Program (AWEP); reauthorizes the innovative grants program (see "EQIP—Conservation Innovation Grants (CIG)"); retains the allocation of 60% of funding each year to practices related to livestock production and reauthorizes and reduces the air quality funding carve- out from $37.5 million to $25 million annually.
National scope	Available nationwide. In FY2017, EQIP obligated over $1.1 billion across 38,726 contracts covering 11.6 million acres.
Leading states	In FY2017, the top three states by active and completed contracts were TX (4,314), MS (2,684), and AL (1,810). The most funding obligated was in TX ($158 million), CA ($105 million), and AR ($62 million).
Backlog/ Interest	In FY2017, 38,726 applications were funded (29.2%), and 56,968 applications went unfunded. The total estimated cost of this backlog is $1.7 billion. The highest numbers of unfunded applications were submitted in MS (6,020), AR (5,853), and OK (3,414).
Funding authority	Mandatory, subject to sequestration. FY2014—$1.35 billion, FY2015—$1.6 billion, FY2016—$1.65 billion, FY2017—$1.65 billion, FY2018—$1.75 billion, and FY2019—$1.75 billion.
FY2018 funding	$1.6 billion (authorization reduced by approximately $115.5 million from sequestration).
FY2019 Administration request	$1.5 billion. Of the proposed $252 million reduction from authorized funding, approximately $116 million is reduced by sequestration and $136 million would be rescinded.
Statutory authority	Authorized in subtitle D of Title III (§§331-336) of the Federal Agriculture Improvement and Reform Act of 1996 (P.L. 104-127) as §§1240-1240G of the Food Security Act of 1985 (P.L. 99-198), as amended. Amended by §§2201-2206 of the Agricultural Act of 2014 (P.L. 113-79). 16 U.S.C. §§3839aa-3839aa-7.
Authorization expires	September 30, 2019.
Program website	http://www.nrcs.usda.gov/wps/portal/nrcs/main/national/programs/financial/eqip

Table 14. EQIP—Conservation Innovation Grants (CIG)

Administering agency(s)	**NRCS**
Program description	CIG is a subprogram of EQIP that awards competitive grants to state and local agencies, nongovernmental organizations, tribes, and individuals to implement innovative conservation techniques and practices. Annual requests for proposals are posted on http://www.grants.gov and include separate funding categories for national and state level competitions. Examples of eligible projects include market systems for pollution reduction, demonstrating precision agriculture, capturing nutrients through a community anaerobic digester, and establishing a tribal partnership for regional habitat conservation.
Major 2014 farm bill amendments	Adds research and demonstration activities, and new technology pilot testing as eligible projects; reauthorizes but reduces the air quality funding carve-out to $25 million of EQIP annually through FY2018; and adds a reporting requirement that no later than December 31, 2014, and every two years thereafter, a report must be submitted to Congress regarding CIG funding, project results, and technology transfer efforts.
National scope	Available nationwide with select states offering state competitions. In FY2017, CIG awarded a total of $22.6 million to 33 projects through the national CIG competition. CIG funding since 2004, when the program began, totals $286.7 million for 711 projects. Applications for FY2018 funding closed February 26, 2018. An estimated $10 million was available for FY2018.
Leading states	None identified.
Backlog/ Interest	None identified.
Funding authority	Unspecified subset of EQIP.
FY2018 funding	Unspecified subset of EQIP.
FY2019 Administration request	Unspecified subset of EQIP, subject to reauthorization.
Statutory authority	Authorized in §2301 of the Farm Security and Rural Investment Act of 2002 (P.L. 107-171) as §1240H of the Food Security Act of 1985 (P.L. 99-198), as amended. Amended by §2207 of the Agricultural Act of 2014 (P.L. 113-79). 16 U.S.C. §3839aa-8.
Authorization expires	September 30, 2018.
Program website	http://www.nrcs.usda.gov/wps/portal/nrcs/main/national/programs/financial/cig

Table 15. Grassroots Source Water Protection Program

Administering agency(s)	**FSA**
Program description	Grassroots Source Water Protection Program provides funding to the National Rural Water Association for technical assistance to operate state's source water protection program. Local programs encourage the voluntary adoption of practices that prevent drinking water pollution.
Major 2014 farm bill amendments	Reauthorizes discretionary funding authority and authorizes $5 million in mandatory funding to remain available until expended.
National scope	In September 2013, the program was expanded to all 50 states.
Leading states	Annual appropriations are divided equally among all states.
Backlog/ Interest	None identified.
Funding authority	Mandatory, subject to sequestration: FY2014—$5 million (to remain available until expended). Discretionary: $20 million annually.
FY2018 funding	$6.5 million (discretionary).
FY2019 Administration request	$0
Statutory authority	Authorized in §2502 of the Farm Security and Rural Investment Act of 2002 (P.L. 107-171) as §1240O of the Food Security Act of 1985 (P.L. 99-198), as amended. Amended by §2502 of the Agricultural Act of 2014 (P.L. 113-79). 16 U.S.C. §3839bb-2
Authorization expires	September, 30, 2018.
Program website	http://www.fsa.usda.gov/programs-and-services/conservation-programs/source-water- protection/index

Table 16. Healthy Forests Reserve Program (HFRP)

Administering agency(s)	**NRCS**
Program description	HFRP assists landowners in restoring and enhancing forest ecosystems using 10-year agreements, 30-year contracts, 30-year easements, and permanent easements.
Major 2014 farm bill amendments	Eliminates mandatory funding authority and replaces it with an authorization to receive appropriations. Adds a definition of "acreage owned by Indian tribes." Provides flexibility for funding technical assistance.

Table 16. (Continued)

Administering agency(s)	**NRCS on private lands and U.S. Forest Service on National Forest Systems lands.**
National scope	Not available nationwide. Limited participation in select states. As of the end of FY2017, 105 agreements covering 676,181 acres have been enrolled in the program.
Leading states	In FY2017, states with the most funding obligated were MS ($109,000), KY ($15,000), and OK ($12,000).
Backlog/ Interest	Not available.
Funding authority	Discretionary. $12 million annually.
FY2018 funding	$0
FY2019 Administration request	$0
Statutory authority	Authorized in Title V Healthy Forest Restoration Act of 2003 (P.L. 108-148), as amended. Amended by §8203 of the Agricultural Act of 2014 (P.L. 113-79). 16 U.S.C. §6572.
Authorization expires	September 30, 2018.
Program website	http://www.nrcs.usda.gov/wps/portal/nrcs/main/national/programs/easements/forests

Table 17. Regional Conservation Partnership Program (RCPP)

Administering agency(s)	**NRCS**
Program description	RCPP provides financial and technical assistance for multi-state or watershed-scale projects. The program creates partnership opportunities to target and leverage federal conservation funding for specific areas and resource concerns. Project areas are defined by eligible partners and are selected through a competitive state or national competition. Partnership agreements are for five years with a possible one-year extension. In addition to defining the project area, providing assistance, and possibly acting on behalf of the producers within the project area, partners must also provide a "significant portion" of the overall cost of the project. Funds are also directed through "critical conservation areas" (CCA) selected by NRCS. Current CCAs include Chesapeake Bay Watershed, Great Lakes Region, Mississippi River Basin, Colorado River Basin, Longleaf Pine Range, Columbia River Basin, Prairie Grasslands Region, and California Bay Delta. Funding is statutorily divided as: critical conservation areas—35%, national projects—40%, and state projects— 25%.

Administering agency(s)	**NRCS**
Major 2014 farm bill amendments	Created in the 2014 farm bill from four repealed programs—Agricultural Water Enhancement Program, the Cooperative Conservation Partnership Initiative, the Chesapeake Bay Watershed Program, and the Great Lakes Basin Program for soil erosion and sediment control. RCPP contracts follow the existing rules and requirements of the covered programs.
National scope	To be eligible for an RCPP contract, a producer must be located in either a CCA or a selected partnership area, but is not required to work with the sponsoring project partner and may choose to work directly with NRCS. Partnership applications are accepted in two phases: pre-proposal and full proposal. In FY2018, 91 projects were selected totaling over $220 million in federal spending.
Leading states	Following the final selection in FY2018, the three funding categories received the following (most total partners; most federal investment): state—54 projects (IL and IN—33 partners; NM—$3.4 million), CCA—19 projects (Mississippi River Basin—66 partners; Chesapeake Bay Watershed—$16.1 million total), and national—18 projects (Driftless Area Habitat for the Wild and Rare Phase 2—46 partners; Crisis to Opportunity: Sierra Nevada Tree Mortality and Partnership GRIC Water Supply Protection Program—$10 million).
Backlog/Interest	In FY2018, the total amount of funding requested was three times the amount available. During the pre-proposal round, 164 proposals were received requesting $683 million with a partner match of $1 billion.
Funding authority	Mandatory, subject to sequestration. RCPP receives 7% of available covered conservation program funds (EQIP, CSP, ACEP, and HFRP) plus an additional $100 million annually.
FY2018 funding	$93.4 million (authorization reduced by approximately $6.6 million from sequestration), plus 7% of available covered conservation program funds (minus sequestration).
FY2019 Administration request	$93.8 million (authorization reduced by approximately $6.2 million from sequestration), plus 7% of available covered conservation program funds (minus sequestration), subject to reauthorization.
Statutory authority	Authorized in subtitle E of Title II (§2401) of the Agricultural Act of 2014 (P.L. 113-79) as §1271 of the Food Security Act of 1985 (P.L. 99-198), as amended. 16 U.S.C. §§3871- 3871f.
Authorization expires	September 30, 2018.
Program website	http://www.nrcs.usda.gov/wps/portal/nrcs/main/national/programs/farmbill/rcpp/

Table 18. Voluntary Public Access and Habitat Incentive Program

Administering agency(s)	**NRCS**
Program description	The Voluntary Public Access and Habitat Incentive Program encourages owners and operators of privately held farm, ranch, and forest land to voluntarily make that land available for access by the public for wildlife-dependent recreation, including hunting or fishing, under programs implemented by state or tribal governments. Competitive grants are offered to states and tribal governments for expanding existing access programs or creating new programs. Grants are reduced by 25% if opening dates for migratory bird hunting in a state are not consistent for residents and nonresidents.
Major 2014 farm bill amendments	Reduces and extends authorization of mandatory funding. Requires USDA to submit a report to Congress no later than 2016 on the effectiveness of the program.
National scope	Available nationwide. In FY2014 and FY2015, $20 million was made available each year. In total, 21 states and one tribal government were awarded agreements. In FY2016, all formal grant agreements were complete.
Leading states	Of the FY2014 and FY2015 selected proposals, states receiving the most funding were PA ($6 million), IA ($3 million), and KS ($2.7 million).
Backlog/ Interest	In FY2014, 28 state wildlife agencies and two tribal governments submitted proposals requesting funding in excess of $62 million. Nine state agencies and one tribal government received $19.7 million. In FY2015, 25 state wildlife agencies submitted proposals requesting funding of over $33 million. Fifteen state agencies received $19.7 million.
Funding authority	Mandatory, subject to sequestration. $40 million in total for the period of FY2014-FY2018.
FY2018 funding	$0
FY2019 Administration request	$0
Statutory authority	Authorized in Title II, §2606 of the Food, Conservation, and Energy Act of 2008 (P.L. 110- 246) as §1240R of the Food Security Act of 1985 (P.L. 99-198), as amended. Amended by §2503 of the Agricultural Act of 2014 (P.L. 113-79). 16 U.S.C. §3839bb-5.
Authorization expires	September 30, 2018.
Program website	http://www.nrcs.usda.gov/wps/portal/nrcs/detail/national/programs/farmbill/?cid= stelprdb1242739

Table 19. Water Bank Program

Administering agency(s)	**NRCS**
Program description	The Water Bank Program offers 10-year, nonrenewable rental agreements to landowners to maintain wetlands in lieu of draining the land for agricultural production. No financial assistance is offered for conservation practices. Applications are ranked based on land use type and flooding impact. Payment rates are as follows: $50/acre/year for cropland, $35/acre/year for pasture and range land (grazing lands), and $20/acre/year for forestland. The program was authorized in 1970 and operated until funding was eliminated in 1994 in favor of longer-term conservation programs. After 17 years of no funding, the program was appropriated $7.5 million in FY2012. These funds were obligated exclusively for agreements in Minnesota, North Dakota, and South Dakota and were focused on flood reduction.
Major 2014 farm bill amendments	None.
National scope	Not available nationwide. Eligible states include MN, ND, and SD. In FY2017, $4.4 million was obligated to 101 agreements covering 11,486 acres.
Leading states	In FY2017, eligible states receiving the most funding were ND ($4 million), SD ($442,000), and MN ($42,000).
Backlog/ Interest	Not available.
Funding authority	Discretionary. No specific authorization level.
FY2018 funding	$4 million.
FY2019 Administration request	$0
Statutory authority	Authorized in the Water Bank Act (P.L. 91-559), as amended. 16 U.S.C. §§1301-1311.
Authorization expires	Permanent authorization.
Program website	http://www.nrcs.usda.gov/wps/portal/nrcs/detail/national/programs/financial/?&cid= stelprdb1047790

Table 20. Watershed and Flood Prevention Operations

Administering agency(s)	**NRCS**
Program description	Also referred to as the Small Watershed Program, the Watershed and Flood Prevention Operations (WFPO) program consists of projects built under two authorities—the Watershed Protection and Flood Prevention Act of 1954 (P.L. 83-566) and the Flood Control Act of 1944 (P.L. 78-534). The vast majority of the projects have been built pursuant to the authority of P.L. 83-566 (referred to as P.L. 566 projects), under which smaller projects authorized by the chief of the NRCS are constructed. Larger projects must be approved by Congress. Eleven projects were specifically authorized under P.L. 78- 534 (referred to as P.L. 534 projects); they are much larger and more expensive than P.L. 566 projects. Under P.L. 566 projects, NRCS provides technical and financial assistance to state and local organizations to plan and install measures to prevent erosion, sedimentation, and flood damage and to conserve, develop, and utilize land and water resources. The project costs are shared with local partners. No project may exceed 250,000 acres, and no structure may exceed more than 12,500 acre-feet of floodwater detention capacity, or 25,000 acre- feet of total capacity. The Senate and House Agriculture Committees must approve projects that need an estimated federal contribution of more than $25 million for construction or include a storage structure with a capacity in excess of 2,500 acre-feet. If the storage structure will have a capacity in excess of 4,000 acre-feet, approval is also required from the Senate Environment and Public Works Committee and the House Transportation and Infrastructure Committee.
Major 2014 farm bill amendments	None.
National scope	Available nationwide. There are 2,100 active or completed P.L. 566 projects in 50 states and U.S. territories. In FY2017, funding was provided to 48 new projects and 41 existing projects.
Leading states	None identified.
Backlog/Interest	Not available.
Funding authority	Discretionary. No specific authorization level.
FY2018 funding	$150 million appropriated, plus $5.6 million in congressionally directed funds from Conservation Operations.
FY2019 Administration request	$0
Statutory authority	Authorized in the Flood Control Act of 1944 (P.L. 78-534), as amended, and the Watershed Protection and Flood Prevention Act (P.L. 83-566), as amended. 33 U.S.C. §701b-1 and 16 U.S.C. §1001 *et. seq.*
Authorization expires	Permanent authorization.
Program website	http://www.nrcs.usda.gov/wps/portal/nrcs/detailfull/national/programs/landscape /?&cid= nrcs143_008271

Table 21. Watershed Rehabilitation Program

Administering agency(s)	**NRCS**
Program description	The Watershed Rehabilitation Program provides technical and financial assistance for planning, design, and implementation to rehabilitate aging watershed dam projects (including upgrading or removing dams) in communities to address health and safety concerns. Only dams constructed under the Watershed and Flood Prevention Operations program are eligible. Small watershed project dams have a 50-year design life. As of the end of 2017, 5,450 of these dams have reached or exceeded that time span. By the end of 2018, this number is expected to rise to 5,922.
Major 2014 farm bill amendments	Reauthorizes both mandatory and discretionary funding authority.
National scope	Only available for dams built through the Watershed and Flood Prevention Operations program and the Resource Conservation and Development (RC&D) program. A total of 284 rehabilitation projects have been authorized in 31 states between FY2000 and FY2017. Of these, a total of 144 projects are complete, and 140 projects are implemented (either in planning, design, or construction phase) subject to funding.
Leading states	States with the largest number of dam projects funded over the life of the program (2000- 2017) are OK (53), TX (35), UT (22), and MS (22). States receiving the most funds in FY2017 were UT ($5.4 million), GA ($4.9 million), and TX ($4.2 million).
Backlog/ Interest	In FY2017 unfunded requests for federal funding totaled over $741.1 million, including 175 dam assessments. The largest total of unfunded requests was submitted in TX ($241 million), MS ($91 million), and WV ($78 million).
Funding authority	Mandatory subject to sequestration: FY2014—$250 million (to remain available until expended). Discretionary: FY2008-FY2018—$85 million annually.
FY2018 funding	$10 million in discretionary funding and $55 million (minus sequestration) in carryover mandatory funding from prior years' reductions.
FY2019 Administration request	$0 in discretionary funding and $0 million in mandatory funding (authorized carryover requested to be reduced by $55 million). Requests a rescission of the proposed reduction ($55 million, minus sequestration).
Statutory authority	Authorized in §313 of the Grain Standards and Warehouse Improvement Act of 2000 (P.L. 106-472) as §14 of the Watershed Protection and Flood Prevention Act, as amended. Amended by §2505 of the Agricultural Act of 2014 (P.L. 113-79). 16 U.S.C §1012.
Authorization expires	September 30, 2018.
Program website	http://www.nrcs.usda.gov/wps/portal/nrcs/main/national/programs/landscape/wr

In: Agricultural Issues
Editor: Michael Butlin

ISBN: 978-1-53615-474-0

Chapter 3

FARM POLICY: USDA'S TRADE AID PACKAGE*

Randy Schnepf, Jim Monke, Megan Stubbs and Jenny Hopkinson

SUMMARY

In early 2018, the Trump Administration—citing concerns over national security and unfair trade practices—imposed increased tariffs on certain imported products in general and on U.S. imports from China in particular. Several of the affected foreign trading partners (including China) responded to the U.S. tariffs with their own retaliatory tariffs targeting various U.S. products, especially agricultural commodities.

On July 24, 2018, Secretary of Agriculture Sonny Perdue announced that the U.S. Department of Agriculture (USDA) would be taking several temporary actions to assist farmers in response to trade damage from what the Administration has characterized as "unjustified retaliation." Specifically, the Secretary said that USDA would authorize up to $12 billion in financial assistance—referred to as a trade aid package—for certain agricultural commodities using Section 5 of the Commodity Credit

* This is an edited, reformatted and augmented version of Congressional Research Service Publication No. R45310, dated Updated January 10, 2019.

Corporation (CCC) Charter Act (15 U.S.C. 714c). USDA intends for the trade aid package to provide short-term assistance in response to the ongoing trade disputes. However, the Secretary stated that there would not be further trade-related financial assistance beyond this $12 billion package. The aid package includes (1) a Market Facilitation Program (MFP) of direct payments (valued at up to $10 billion) to producers of soybeans, corn, cotton, sorghum, wheat, hogs, and dairy who are most affected by the trade retaliation (sweet cherries and almonds were added to this list in September); (2) a Food Purchase and Distribution Program to partially offset lost export sales of affected commodities ($1.2 billion); and (3) an Agricultural Trade Promotion (ATP) Program to expand foreign markets ($200 million).

USDA's Farm Service Agency will administer the MFP by providing payments in two potential tranches: a first round announced on August 27, 2018, initially valued at $4.7 billion; and an equivalent-valued second round announced on December 17, 2018. However, producers need only sign up once for the MFP to be eligible for first and second payments. The sign-up period for soybeans, corn, cotton, sorghum, wheat, hogs, and dairy started September 4, 2018. The sign-up period for fresh sweet cherries and shelled almonds started on September 24. To be eligible, a producer must have an ownership share in the commodity, be actively engaged in farming, and be in compliance with adjusted gross income restrictions and conservation provisions. Eligible producers should apply after their harvest is complete. Initially, producers were given a deadline of January 15, 2019, to complete an application. However, USDA extended the deadline on January 8, to an indefinite time period due to the government shutdown.

USDA used 2017 production data to estimate that approximately $9.6 billion would be distributed in MFP payments for corn, cotton, sorghum, soybeans, wheat, dairy, hogs, fresh sweet cherries, and shelled almonds, with over three-fourths ($7.3 billion) of MFP payments provided to soybean producers. MFP payments are capped on a per-person or per-legal-entity basis at a combined $125,000 for eligible crop commodities, a combined $125,000 for dairy production and hogs, and, separately, a combined $125,000 for fresh sweet cherries and shelled almonds.

In addition to the MFP payments, the Administration announced a Food Purchase and Distribution Program that is to undertake $1.2 billion in government purchases of excess food supplies. USDA has targeted an initial 29 commodities for purchases and distribution through domestic nutrition assistance programs. Purchasing orders and distribution activities are to be adjusted based on the demand by the recipient food assistance programs geographically.

The smallest piece of the trade aid package is an allocation of $200 million to the ATP to boost the trade promotion efforts at USDA's

Foreign Agricultural Service, including foreign market development for affected agricultural products.

USDA's use of its discretionary authority under the CCC Charter Act to make direct payments without further congressional action has historically been somewhat intermittent and limited in its scale. While the use of this authority is not without precedent, the scope and scale of this trade aid package has increased congressional and public interest. Furthermore, the significant variation in the announced MFP payment rates for affected commodities has elicited questions about equitable treatment among affected commodities. On September 13, USDA released a description of its MFP payment methodology, which is based strictly on the estimated direct trade "damage"—that is, export losses resulting from retaliatory tariffs. Indirect effects—such as the decline in market prices and resultant "lost value" for many of the affected commodities—were not included in the payment calculation.

INTRODUCTION

In early 2018, the Trump Administration—citing concerns over national security and unfair trade practices—imposed increased tariffs on certain imported products in general and on U.S. imports from China in particular.[1] Several of the affected foreign trading partners responded to the U.S. tariffs with their own retaliatory tariffs targeting various U.S. products, especially agricultural commodities.[2]

On July 24, 2018, Secretary of Agriculture Sonny Perdue announced that the U.S. Department of Agriculture (USDA) would be taking several temporary actions to assist farmers in response to trade damage from what the Administration has characterized as "unjustified retaliation."[3] Specifically, USDA would authorize up to $12 billion in financial assistance—referred to as the "trade aid" package—for certain agricultural commodities under Section 5 of the Commodity Credit Corporation (CCC) Charter Act (15 U.S.C 714c). The Secretary said that most of the funding would go to agricultural commodities most directly affected by the trade

[1] See CRS Insight IN10943, *Escalating Tariffs: Timeline*.

[2] CRS Insight IN10880, *China's Retaliatory Tariffs on Selected U.S. Agricultural Products*.

[3] USDA, "USDA Assists Farmers Impacted by Unjustified Retaliation," press release, July 24, 2018.

retaliation— corn, cotton, soybeans, sorghum, wheat, hogs, and dairy (sweet cherries and almonds were added to this list in September)—but that some funding would also be used for the purchase, distribution, and trade promotion of a variety of affected commodities.

The trade-aid package includes a Market Facilitation Program (MFP) of direct payments to affected producers, a Food Purchase and Distribution Program, and an Agricultural Trade Promotion (ATP) program.[4] Payments under the MFP program would be made in two rounds: a first round announced on August 27, 2018, initially valued at $4.7 billion; and an equivalent-valued second round announced on December 17, 2018.[5] Secretary Perdue stated that there would not be further trade-related financial assistance beyond this $12 billion package as producers would be able to adjust their production activities in 2019 to reflect market conditions related to the trade dispute.

This chapter provides background on the trade dispute that triggered the trade-aid package as well as the authority used by USDA to respond to the trade dispute with financial assistance. Then the report describes the three components of the trade-aid package with details on their implementation.

TARIFFS AS THE ORIGIN OF THE TRADE AID PACKAGE

In March 2018, the Trump Administration began applying a 25% tariff to U.S. steel imports and 10% tariff to U.S. aluminum imports from certain countries, citing national security concerns.[6] In April, in response to alleged unfair trade practices by the Chinese government, the Administration placed additional tariffs on a number of Chinese products

---

[4] USDA, "USDA Announces Details of Assistance for Farmers Impacted by Unjustified Retaliation," press release, August 27, 2018.

[5] USDA, "USDA Launches Second Round of Trade Mitigation Payments," press release, December 17, 2018, https://www.usda.gov/media/press-releases/2018/12/17/usda-launches-second-round-trade-mitigation-payments.

[6] CRS Report R45249, *Section 232 Investigations: Overview and Issues for Congress*.

that are exported to the United States.[7] China, Canada, Mexico, the European Union, and Turkey subsequently enacted retaliatory tariffs on U.S. food and agricultural products, in addition to other goods, in response to the U.S. actions.[8] The retaliatory tariffs from those countries now apply to more than 800 U.S. food and agricultural products across meats, grains, dairy products, specialty and horticultural crops, seafood, and alcoholic beverages. The export value for the targeted products to the retaliating countries totaled about $26.9 billion in 2017—about 18% of total U.S. agricultural exports.[9]

China, which is subject to the largest set of U.S. tariff increases—including both the U.S. steel and aluminum tariffs and the U.S. tariffs in response to unfair trade practices—also has the most expansive list of retaliatory tariffs. All told, China, which was the second-leading export market by value for U.S. food and agriculture products in 2017, has levied retaliatory tariffs on about 800 U.S. food and agricultural products that were worth about $20.6 billion in exports to that country in 2017.[10] Among China's retaliatory tariffs is a 25% tariff on soybeans, its top agricultural product import by value from the United States. China imported about $12 billion worth of U.S. soybeans in 2017, accounting for 57% of the value of all U.S. soybean exports that year. With the higher tariffs in place, China is now purchasing more soybeans from Brazil and elsewhere to meet its demand.[11] China has also targeted other key U.S. products, including sorghum, wheat, pork and pork offal, dairy products, fruits and nuts, seafood, and whiskey.

---

[7] CRS In Focus IF10708, Enforcing U.S. Trade Laws: Section 301 and China.

[8] Agriculture and food products covered in this report include most of chapters 1-24 of the U.S. Harmonized Tariff Schedule (HTS), which cover meat, grains, animal feed, dairy, horticultural products, processed food, unprocessed tobacco, seafood, and alcoholic beverages. This list also includes essential oils (HTS chapter 33), animal hides and skins (chapters 41 and 43), and cotton and wool (chapters 51 and 52). The harmonized schedule is a hierarchical structure for describing all goods in trade for duty, quota, and statistical purposes.

[9] USDA Global Agricultural Trading System, August 20, 2018.

[10] USDA Foreign Agricultural Service (FAS). *China Announces Supplemental Tariffs in Response to U.S. 301 Tariffs*, GAIN Report CH18043, August 6, 2018.

[11] Karl Plume, "U.S. Soybean Exports Scrapped as China Shifts to Brazilian Beans," Reuters, May 18, 2018.

Among other countries, Canada—the leading export market for U.S. agriculture and food products in 2017—has imposed retaliatory tariffs of 10% on about 20 food and agricultural products, mostly processed foods.[12] U.S. exports of those products to Canada in 2017 were valued at $2.6 billion. Mexico, the third-leading export market for U.S. agriculture and food products by value in 2017, has imposed tariffs ranging from 15% to 25% on cheese, pork, and some prepared foods.[13] U.S. exports of those products to Mexico were valued at about $2.5 billion in 2017. The European Union has levied tariffs on a small number of U.S. prepared foods, corn, and rice, which were worth about $1 billion in 2017.[14] Turkey has imposed retaliatory tariffs on U.S. nuts, rice, and some prepared foods, imports of which amounted to some $250 million in 2017.[15]

U.S. agriculture and food products have been targeted with increased tariffs by foreign nations for several reasons. First, the United States exports a large amount of agriculture and food products, so many countries have the choice of retaliating against those goods. Second, agricultural commodities are easily substituted from among potential suppliers, so curbing imports from one country would not necessarily limit an importing country's access to the commodity. For example, China has turned primarily to Brazil for more of its soybean imports. Third, given the geographic nature of the production of some agriculture and food products, countries can target certain goods in order to negatively and disproportionately affect the constituents of specific U.S. lawmakers. For example, all of the retaliating countries have imposed retaliatory tariffs on whiskey, some specifically on Bourbon whiskey, which is largely

---

[12] For the full list of products subject to Canadian retaliatory tariffs, see Canada Department of Finance, "Notice of Intent to Impose Countermeasures Action Against the United States in Response to Tariffs on Canadian Steel and Aluminum Products," https://www.fin.gc.ca/activty/consult/cacsap-cmpcaa-eng.asp.

[13] FAS, *Mexico Announces Retaliatory Tariffs*, GAIN Report MX8028, June 6, 2018.

[14] FAS, *EU Imposes Additional Tariffs on U.S. Products*, GAIN Report E18045, June 21, 2018.

[15] FAS, *Turkey Introduces New Additional Levy on U.S. Products*, June 28, 2018.

produced in Kentucky, rather than on all distilled beverages or alcohol more generally.[16]

Trade Aid Package Authority

The primary authority for the trade aid package is the Secretary of Agriculture's discretion to use the general powers of the CCC. The CCC is a wholly government-owned entity that exists solely to finance authorized programs that support U.S. agriculture. It is federally chartered by the CCC Charter Act of 1948 (P.L. 80-806; 15 U.S.C. 714 *et seq.*), as amended. Most CCC-funded programs are classified as mandatory spending programs and therefore do not require annual discretionary appropriations in order to operate.[17] The CCC instead borrows from the U.S. Treasury to finance its programs consistent with its permanent, indefinite authority to borrow up to $30 billion. Congress replenishes the CCC borrowing authority by appropriating funding to cover the CCC's net realized losses.[18]

Typically, Congress passes laws, such as omnibus farm bills, that specifically direct USDA on how to administer CCC activities and in what amounts to fund them. The underlying authorization for the CCC, however, also provides the Secretary with general powers to take certain actions in support of U.S. agriculture at the discretion of the Secretary. This discretionary use has historically been somewhat intermittent and limited in its scale, but it is the basis of the MFP and ATP announced by the Administration.[19]

---

[16] Amanda Macias, "America's Booming Bourbon Business Caught in the Crosshairs of Trump's Trade War," *CNBC*, June 17, 2018.

[17] CRS In Focus IF10783, Farm Bill Primer: Budget Issues.

[18] For more detailed information on the CCC, see CRS Report R44606, The Commodity Credit Corporation: In Brief; or CRS Insight IN10941, Commodity Credit Corporation: Q&A.

[19] CCC, "Market Facilitation Program," 83 Federal Register 44173, August 30, 2018; and CCC, "Agricultural Trade Promotion Program," 83 Federal Register 44178, August 30, 2018.

USDA also has discretionary authority to purchase U.S. agricultural commodities under a provision known as Section 32.[20] The name refers to its authorization in Section 32 of the act of August 24, 1935 (P.L. 74-320; 7 U.S.C. 612c), as amended.[21] Most of Section 32's mandatory funding is transferred to the USDA's child nutrition account, but the Secretary has broad discretion in how to spend the remaining unallocated funding—some of which is used to purchase agricultural commodities. The premise is that removing products from normal marketing channels helps to reduce supply and thereby increase prices and farm income. Purchased commodities are diverted to domestic food assistance programs as discussed below (see "Food Purchase and Distribution Program").

The Administration's trade aid announcement does not specify whether the CCC or Section 32 authority is being used to make the purchases under the announced Food Purchase and Distribution Program. However, the scale of the $1.2 billion program indicates that the CCC is most likely the source since the typical annual amount of funding available in Section 32 for purchases is rarely more than half of this amount. Whether from the CCC or Section 32, the Administration's purchases appear to use distribution channels similar to those under Section 32.

TRADE AID PACKAGE IMPLEMENTATION

On August 27, 2018, Secretary Perdue announced the first round of trade assistance. As part of the August 27 announcement, Secretary Perdue provided details on each of the three trade aid package components, including an initial tranche of $6.1 billion in designated outlays out of a potential $12 billion in total program spending. The MFP was to provide initial estimated direct payments of $4.7 billion to qualifying agricultural producers. A Food Purchase and Distribution Program is to undertake $1.2 billion in government purchases of excess food supplies. The ATP

[20] For additional information, see CRS Report RL34081, Farm and Food Support Under USDA's Section 32 Program.

[21] It is also referred to as Funds for Strengthening Markets, Income and Supply program.

program, funded with an additional $200 million, is to help finance foreign market development for affected agricultural products. On December 17, 2018, Secretary Perdue revised the first round of MFP outlays upward slightly to $4.8 billion, and announced an equivalent $4.8 billion in potential second-round outlays.

Market Facilitation Program

The MFP[22] provides direct financial assistance to producers of commodities that are significantly impacted by actions of foreign governments resulting in the loss of traditional exports. USDA initially determined that qualifying commodities include corn, upland cotton, extra-long-staple cotton, sorghum, soybeans, wheat, dairy, and hogs. On September 21, 2018, USDA announced that fresh sweet cherries and shelled almonds are also eligible for MFP payments.

USDA's Farm Service Agency (FSA) is to administer the MFP by providing payments in two potential tranches.[23] However, producers need only sign up once for the MFP to be eligible for first and second payments.[24] Under the sign-up period, producers can submit MFP applications beginning on the following dates: September 4, 2018, for producers of soybeans, sorghum, corn, wheat, cotton, dairy, and hogs; and September 24, 2018, for producers of shelled almonds and fresh sweet cherries. Eligible producers should apply after their harvest is complete. Initially, producers were given a deadline of January 15, 2019, to complete an application. However, USDA extended the deadline on January 8 until after the shutdown ends.[25] The current deadline for producers to certify their 2018 production is May 1, 2019.

---

[22] The Market Facilitation Program moniker—used to describe the direct payment portion of the trade aid package—is newly created by USDA and does not represent an existing program.

[23] USDA, "Market Facilitation Program (MFP)," August 2018.

[24] USDA, "USDA Launches Second Round of Trade Mitigation Payments," press release, December 17, 2018, https://www.usda.gov/media/press-releases/2018/12/17/usda-launches-second-round-trade-mitigation-payments.

[25] USDA Office of Communication, press release, January 8, 2019. FSA offices were closed starting on December 28, 2018, due to a lack of funding, thus preventing farmers from

USDA used 2017 production data to estimate that approximately $9.6 billion would be distributed in MFP payments for corn, cotton, sorghum, soybeans, wheat, dairy, hogs, fresh sweet cherries, and shelled almonds, with over three-fourths ($7.3 billion) of MFP payments provided to soybean producers (*Table 1*).[26]

Who Qualifies for a Payment?

U.S. producers of corn, cotton, sorghum, soybeans, wheat, dairy, hogs, fresh sweet cherries, and shelled almonds are eligible for MFP payments at this time. Eligible applicants must

- have an ownership interest in the commodity[27] and be actively engaged;[28]
- have an average adjusted gross income for tax years 2014, 2015, and 2016 of less than $900,000 per year;[29]
- comply with the provisions of the "Highly Erodible Land and Wetland
- Conservation" regulations, often called the conservation compliance provisions.[30]

USDA Determination of MFP Per-Unit Payment Rates

USDA determined MFP payments based on its estimated "direct trade damage"—that is, the difference in expected trade value for each affected commodity with and without the retaliatory tariffs (*Table A-3*).[31] The

---

applying or certifying their 2018 production. In the January 8 press release, USDA stated that the deadline for application would be extended by the number of business days that FSA offices are closed as a result of the shutdown.

[26] USDA, "USDA Launches Second Round of Trade Mitigation Payments."

[27] With respect to cotton, ownership applies only to cotton harvested as lint. For corn, sorghum, and wheat, ownership applies only to the portion harvested as grain. For milk, ownership applies only to dairy operations in business as of June 1, 2018, and ownership of hogs is as of August 1, 2018, but excludes hogs produced under contract.

[28] See CRS Report R44656, *USDA's Actively Engaged in Farming (AEF) Requirement*.

[29] See CRS Report R44739, *U.S. Farm Program Eligibility and Payment Limits*.

[30] See CRS Report R42459, *Conservation Compliance and U.S. Farm Policy*.

[31] USDA, "USDA Releases Details Trade Damage Estimate Calculations," press release, September 13, 2018; and USDA, Office of the Chief Economist, "Trade Damage

estimated "trade damage" for each affected commodity was then divided by the crop's production in 2017 to derive a per-unit payment rate. Indirect effects—such as any decline in market prices and resultant "lost value" for many of the affected commodities— are not included in the payment calculation (see *Appendix B*).

USDA's trade-aid package is thus linking MFP commodity payments only to the trade loss associated with each identified MFP commodity. Neither final trade effect, with or without retaliatory tariffs, is observable because much of the affected agricultural production had yet to be harvested and sold at the time the payment rates were calculated, and markets had yet to fully adjust to whatever new trade patterns would emerge from the trade dispute. As a result, USDA estimated both export values (with and without retaliatory tariffs) using a global trade model that took into account the availability of substitute supplies from export competitors, and the availability of demand for U.S. agricultural exports from alternate importers.

How Will Farm-Level MFP Payments Be Determined?

MFP payments are tied directly to a producer's actual level of production of eligible commodities in 2018. A producer's total potential MFP payment for an eligible commodity equals the announced payment rate per unit (see column two of *Table 1*) times the harvested (and certified) production during 2018 or in the case of hogs, the inventory during the period of July 15 to August 15, 2018. During the first payment period (announced by USDA on September 27), MFP payments were set equal to the announced MFP payment rate times 50% of a producer's harvested (and certified) production. The second payment rate (announced on December 17) applied to the remaining 50% of the producer's production.

Estimation for the Market Facilitation Program and Food Purchase and Distribution Program," September 13, 2018.

Table 1. MFP: Eligible Commodities, Payment Rates, and Production Base

Commodity[a]	Per-Unit Rate ($)[b]	Unit	Production Base[c]	Estimated Total Payment ($ Million)[d]
Soybeans	1.65	bushel	Harvested crop in 2018	$7,259.4
Hogs	8.00	head	Inventory during July 15 to August 15, 2018[e]	$580.6
Cotton[f]	0.06	pound	Harvested crop in 2018	$553.8
Sorghum	0.86	bushel	Harvested crop in 2018	$313.6
Dairy (milk)	0.12	cwt.[g]	Historical MPP-Dairy production[h]	$254.8
Wheat	0.14	bushel	Harvested crop in 2018	$238.4
Corn	0.01	bushel	Harvested crop in 2018	$192.0
Fresh sweet cherries	0.16	pound	Harvested crop in 2018[i]	$111.5
Shelled almonds	0.03	pound	Harvested crop in 2018[j]	$63.3
Total				$9,567.4

Source: USDA, "USDA Launches Second Round of Trade Mitigation Payments," press release, December 17, 2018; USDA, "USDA Announces Details of Assistance for Farmers Impacted by Unjustified Retaliation," press release, August 27, 2018; USDA, "Market Facilitation Program (MFP)," August 2018; USDA, Notice of Funds Available, Market Facilitation Program (MFP), August 27, 2018, and "USDA Adds Shelled Almonds and Fresh Sweet Cherry to Market Facilitation Program," press release, September 21, 2018.

Notes: NA = Not available.

[a] Crops that are grazed in the field or used as forage are not eligible for MFP payments.

[b] The per-unit payment rate is based on the USDA-determined trade damage.

[c] A crop producer requesting an MFP payment must have a crop acreage report (Form FSA-578) on file with USDA. Producers who do not have an acreage report would follow the "late-filed" acreage report process.

[d] The estimated total payment is based on the announced MFP per-unit payment rate (column two) times the harvested production base (using 2017 production) or inventory (hogs). This could be received by a producer as a single payment or in two equal tranches.

[e] Payment for hog operations is based on the total number of head of live hogs during the July 15 to August 15, 2018, period that correctly reflects the farm's operations. Production records for hogs may include, but are not limited to, breeding records, inventory records, sales receipts, rendering receipts, or veterinary records.

[f] Both upland cotton and extra-long-staple cotton are eligible for MFP payments.

[g] cwt. = hundred pounds or hundredweight.

[h] The payment for dairy production is based on the historical production reported for the Margin Protection Program for Dairy. For existing dairy operations, the production history is established using the highest annual milk production marketed during the full calendar years of 2011, 2012, and 2013. Dairy operations are also required to have been in operation on June 1, 2018.

[i] Sweet cherries intended for process market or juice are not eligible for MFP. The quantity of production for sweet cherries is on a "pack-out" basis.

[j] Shelled almonds will be based on the total eligible kernels or such similar term as edible meat weight.

The MFP is separate from and in addition to the current safety net support provided by the Agricultural Risk Coverage (ARC) and Price Loss Coverage (PLC) support programs[32] or crop insurance coverage where revenue insurance protects against low prices, low yields, or a combination of both.[33] Furthermore, by coupling the payments directly to production,[34] those regions of the country where drought or other yield-reducing factors have negatively impacted production during 2018 may receive less aid through MFP than other regions.

According to USDA, as of January 9, 2019, about $5.2 billion in payments have been made to farmers, including over 360,000 payments totaling nearly $3 billion since the government shutdown began on December 21, 2018.[35] FSA offices closed on December 28, 2018, due to a lack of funding under the government shutdown. Producers who have not yet applied for payments or certified their 2018 production must wait for FSA offices to reopen before receiving MFP payments. However, USDA has said that producers that have already applied and certified their 2018 production will continue to receive MFP payments during the government shutdown.[36]

MFP Payment Limit

USDA announced that MFP payments are capped on a per-person or per-legal-entity basis under three separate payment limits: a combined $125,000 for eligible crops (corn, cotton, sorghum, soybeans, and wheat), a combined $125,000 for livestock (dairy production and hogs), and a combined $125,000 for eligible specialty crops (fresh sweet cherries and shelled almonds). Furthermore, MFP payments do not count against other 2014 farm bill payment limitations. There are no criteria in place to calculate whether losses covered under revenue support programs (e.g.,

---

[32] CRS In Focus IF10711, Farm Bill Primer: ARC and PLC Support Programs.

[33] See CRS In Focus IF10638, Farm Bill Primer: The Farm Safety Net; or CRS Report R45193, Federal Crop Insurance: Program Overview for the 115th Congress.

[34] CRS Report R43817, 2014 Farm Bill Provisions and WTO Compliance.

[35] Annie Gowen, Jeff Stein and Sean Sullivan, "Already Reeling from Tariff War, Some Farmers Aren't Receiving Government Support Checks Amid Shutdown," *Washington Post*, January 9, 2019.

[36] *Politico*, "Perdue Extends Trade-Aid Deadline," January 9, 2019.

ARC and PLC) of the 2014 farm bill might be duplicated by MFP. As a result, the same program acres that are eligible for ARC or PLC payments may be eligible for MFP payments.

WTO Compliance of Trade Retaliation Assistance Payments

Due to its potential price tag ($12 billion) and the coupled nature of the MFP payments, there is considerable interest from policymakers and market observers about whether these payments will be fully compliant with World Trade Organization (WTO) commitments.[37] It would appear that, if the United States restricts MFP payments to $12 billion or less, and its other amber box payments adhere to the recent annual average of $4.9 billion since 2010, then total U.S. amber box payments would be below its $19.1 billion limit on trade-distorting farm subsidies at the WTO.[38] However, several economists have suggested there is considerable uncertainty in how much the eventual MFP payments will be. For example, Darci Vetter, former chief U.S. agricultural negotiator at the Office of the U.S. Trade Representative, said that current low agricultural commodity prices cause her to worry that billions of dollars in "additional payments will put us over our [amber box] $19 billion cap," exposing the United States to a potential legal challenge.[39] Joe Glauber, a former USDA chief economist, stated, "I would be very hesitant to say categorically, 'No, we're not going to hit our $19.1 [billion ceiling].'"

Industry Response to MFP Payment Allocation

While soybean growers and most farm-advocacy groups have generally been supportive of the payments, some commodity groups—most notably associations representing corn, wheat, and milk—contend that the MFP payments are insufficient to fully compensate their industries (see *Table A-4* and *Appendix B* for a comparison of "trade loss" and

---

[37] See CRS Insight IN10940, Potential WTO Implications of USDA's Proposed Response to Trade-Retaliation.

[38] See CRS In Focus IF10192, WTO Disciplines of Domestic Support for Agriculture.

[39] Doug Palmer, "Trump's Trade Aid Plan Could Breach WTO Farm Subsidy Limit," Politico, August 9, 2018

"market loss").[40] The National Corn Growers Association claims that recent trade disputes have lowered corn prices by $0.44/bu. for a loss of $6.3 billion on the projected 2018 harvest. Similarly, the National Association of Wheat Growers estimates that a $0.75/bu. price decrease will result in nearly $2.5 billion in lost value, while the National Milk Producers Federation calculates that milk prices are now estimated to be $1.10/cwt. lower than just prior to the trade retaliation, causing over $1.2 billion in losses based on milk futures prices.

Many specialty crop groups similarly contend that their interests are not being fully compensated for tariff-related export losses by the USDA trade aid programs. For example, a recent study suggests that, in California alone, specialty crops may suffer trade-related losses of over $3.3 billion this year.[41]

Food Purchase and Distribution Program

The Administration is allocating about $1.2 billion of its trade aid package to purchasing various agricultural commodities and distributing them through domestic nutrition assistance programs. USDA typically purchases agricultural commodities for domestic distribution in two ways: (1) "entitlement purchases" for the mandated, preplanned needs of a feeding program; and (2) "contingency purchases" (also called "bonus buys") that are usually triggered as a surplus removal mechanism to raise market prices of a commodity without displacing normal demand.

The new $1.2 billion of purchases is under the second category of contingency purchases. Contingency purchases are statutorily authorized under the Secretary's discretion to support agriculture by making

[40] *The Hagstrom Report*, "Summary of Trump Trade Aid: It's Not Enough," vol. 8, no. 201 (August 28, 2018); and *Politico*, "Trump Offers Trade Aid to Farmers, but Some Question Its Fairness," August 28, 2018. See also Y. Zhou et al., "Dispatches from the Trade Wars," *farmdoc daily*, August 29, 2018.

[41] Daniel A. Sumner and Tristan M. Hanon, "Economic Impacts of Increased Tariffs That Have Reduced Import Access for U.S. Fruit and Tree Nuts Exports to Important Markets," University of California, August 1, 2018.

purchases under the CCC or Section 32 as discussed above.[42] These are mandatory funds and do not need to be appropriated.

When USDA purchases commodities, especially for distribution to nutrition assistance programs, the Agricultural Marketing Service (AMS) announces its purchasing intentions with product specifications. Vendors who are approved to sell to USDA may submit offers.[43] The purchased products would be distributed through regular USDA nutrition assistance channels that provide in-kind assistance, such as food banks participating in the Emergency Food Assistance Program, the Commodity Supplemental Food Program, child nutrition programs such as the National School Lunch Program, and the Food Distribution Program on Indian Reservations.[44] However, not all of these programs have the authority to accept contingency/bonus purchases.

The Administration's August 27 announcement listed 29 commodities targeted for purchases totaling $1.2 billion (*Table A-1*).[45] It also mentions two additional commodities (sweet cherries and almonds) that total $175 million, with program details to be determined (*Table A-2*). The announced purchase values were set for each affected commodity using the same gross trade damage formula that was used to calculate the MFP per-unit payment rate described earlier.[46]

The largest purchases that were announced include pork ($559 million), apples ($93 million), dairy ($85 million), and pistachios ($85 million). USDA said that the breadth of commodities and scale of purchases was based on economic analyses of the effect of tariffs. Purchasing orders and distribution activities are to be adjusted based on the demand by the recipient food assistance programs geographically. As of December 17, 2018, USDA had procured some portion of 16 of the 29

[42] See CRS Insight IN10941, Commodity Credit Corporation: Q&A; and CRS Report RL34081, Farm and Food Support Under USDA's Section 32 Program.

[43] See AMS, "How the Process Works," https://www.ams.usda.gov/selling-food/how-process-works.

[44] CRS Report R42353, Domestic Food Assistance: Summary of Programs.

[45] USDA's announced purchase values were set at the level of gross trade damage to each affected commodity using the same global trade model as used for calculating the MFP per-unit payments.

[46] Office of the Chief Economist, "Trade Damage Estimation."

commodities included in the program, totaling more than 4,500 truckloads of food. USDA's AMS will continue purchasing commodities for delivery throughout 2019.[47]

In FY2017, the AMS purchased $2.2 billion of commodities for distribution for domestic nutrition assistance.[48] Of this total, $735 million was from Section 32 ($270 million in contingency purchases that are most similar to those under the trade aid package and $465 million in entitlement purchases), and $1.5 billion was entitlement purchases from the USDA's Food and Nutrition Service budget. No purchases were made with CCC funds. Thus, the new program of contingency purchases is several times larger than a typical annual amount and a relatively large increase in the amount distributed through nutrition programs.

Agricultural Trade Promotion Program

The third and smallest element of the trade aid package is the ATP program. The Administration is allocating $200 million of the trade aid package to boost trade promotion efforts of USDA's Foreign Agricultural Service (FAS). The program is to operate in a manner similar to FAS's Market Access Program (MAP) and Foreign Market Development Program (FMDP). These funds are to provide cost-share assistance to eligible U.S. agricultural organizations to promote U.S. food and agricultural goods overseas and develop new markets to help offset the adverse effects of the retaliatory tariffs.[49] The money—which would nearly double the amounts made available annually for the MAP and FMDP trade promotion programs for one year—can be used for such activities as consumer advertising, public relations, point-of-sale demonstrations, participation in trade fairs and exhibits, market research, and technical

---

[47] USDA, "USDA Launches Second Round of Trade Mitigation Payments," press release, December 17, 2018, https://www.usda.gov/media/press-releases/2018/12/17/usda-launches-second-round-trade-mitigation-payments.

[48] USDA, *FY2019 Budget Explanatory Notes for Committee on Appropriations*, "Agricultural Marketing Service," p. 92, and "Commodity Credit Corporation," p. 21.

[49] MAP and FMDP are currently authorized at $234 million annually. See CRS Report R44985, *USDA Export Market Development and Export Credit Programs: Selected Issues*.

assistance. Further, ATP money is not limited to certain commodities and is to be available to all sectors of agriculture.

While the $200 million for ATP is considerably less than the other programs in the trade aid package, it is a notable increase for USDA's trade promotion programs, which are authorized at $234 million annually. Though all sectors of agriculture can apply for ATP funding through eligible U.S. organizations, it is unclear whether USDA intends to give preference to certain commodities—such as those that are not eligible for other programs under the trade aid package or those most impacted by the tariffs.

The application period for ATP closed in November 2018 with more than $600 million in requested activities from more than 70 organizations. USDA's FAS will announce ATP funding awards in early January 2019.[50]

CONCLUSION

The broad discretionary authority granted to the Secretary under the CCC Charter Act to implement the trade aid package also allows the Secretary to determine how the aid is to be calculated and distributed. Using this authority is not without precedent, but the scope and scale of its use for the trade aid package has increased congressional and public interest. USDA has declared this trade aid package to be a temporary, one-time response to foreign tariffs imposed on selected U.S. commodities.

Most farm commodity and advocacy groups have been supportive of the trade aid package even as they have called for solutions that restore export activity. However, some stakeholders have begun to question the equity of the distribution of MFP payments due to difficulties in isolating specific market effects and the initial lack of transparency around the formulas for determining MFP payment rates. Now that the formulas are public, several commodity groups question the rationale for determining MFP payments based on "trade damage" rather than the broader "market

---

[50] See CRS Report R44985, *USDA Export Market Development and Export Credit Programs: Selected Issues.*

loss" measure. Some trade economists and market watchers have suggested that the potential effects of the trade aid package and the imposition of tariffs and retaliatory tariffs could be longer lasting because they have created uncertainty about U.S. trade policy behavior and have called into question U.S. reliability as a trading partner.[51] Further, the use of CCC authority to mitigate tariff-related losses may establish a precedent for future situations.

APPENDIX A. FOOD PURCHASES IN THE TRADE AID PACKAGE

Table A-1. USDA Trade Aid Package Food Purchases

Commodity	Target Amount ($1,000s)
Pork	$558,800
Apples	$93,400
Pistachios	$85,200
Dairy	$84,900
Oranges	$55,600
Grapes	$48,200
Rice	$48,100
Potatoes	$44,500
Walnuts	$34,600
Cranberries	$32,800
Orange Juice	$24,000
Plums/Prunes	$18,700
Navy Beans	$18,000
Pecans	$16,000
Beef	$14,800
Kidney Beans	$14,200
Peanut Butter	$12,300
Peas	$11,800
Macadamia	$7,700
Lemons/Limes	$3,400

[51] Mario Parker, Isis Almeida and Alix Steel, "Cargill CEO Sees Risk to U.S. Farmers as China Shuns Soybeans," *Bloomberg News*, September 25, 2018, https://www.bloomberg.com/news/articles/2018-09-25/cargill-ceo-sees-longterm-risk-to-farmers-in-u-s-china-spat.

Table A-1. (Continued)

Commodity	Target Amount ($1,000s)
Sweet Corn	$2,400
Hazelnuts	$2,100
Lentils	$1,800
Blueberries	$1,700
Strawberries	$1,500
Pears	$1,400
Grapefruit	$700
Apricots	$200
Figs	$15
Total	$1,238,815

Source: USDA, "USDA Announces Details of Assistance for Farmers Impacted by Unjustified Retaliation," press release, August 27, 2018.

Table A-2. USDA Trade Aid Package Food Purchases, Details to Be Determined

Commodity	Target Amount ($1,000s)
Almonds	$63,300
Sweet Cherries	$111,500
Total	$174,800

Source: USDA, "USDA Announces Details of Assistance for Farmers Impacted by Unjustified Retaliation," press release, August 27, 2018.

Table A-3. Trade-Affected Share of Production for Affected Commodities

Commodity	Unit Retaliation	2017 Production	2017 Exports Facing	Share (%)
Soybeans	Million bu.	4,392.0	1,166.0	27%
Hogs	Million head	25,598.0	3,705.6	14%
Cotton	Million lbs.	10,042.8	1,175.8	12%
Sorghum	Million bu.	364.0	181.4	50%
Dairy (milk)	Million cwt.	2,155.0	46.3	2%
Wheat	Million bu.	1,741.0	55.6	3%
Corn	Million bu.	4,604.0	58.9	0%
Fresh sweet cherries	Million lbs.	875.1	NA	NA
Shelled almonds	Million lbs.	2,270.0	NA	NA

Source: USDA, "Cost Benefit Analysis—Market Facilitation Program," July 24, 2018.

Notes: Data for the year 2017 is most current data available at time of publication of source.

Table A-4. Comparison of USDA Commodity Price Forecasts (May to September, 2018)

	2018 Farm Price Forecast[a]			Difference	
Commodity	Unit	May	September	Value	% Change
Soybeans	$/bushel	10.50	8.60	-1.90	-18.1%
Hogs	$/cwt.	44.50	40.50	-4.00	-9.0%
Commodity	Unit	May	September	Value	% Change
Sorghum	$/bushel	3.60	3.30	-0.30	-8.3%
Corn	$/bushel	3.80	3.50	-0.30	-7.9%
Wheat	$/bushel.	5.00	5.10	+0.10	+2.0%
Dairy (milk)	$/cwt.	16.75	17.25	+0.50	+3.0%
Cotton	$/lb.	0.65	0.75	+0.10	+15.4%

Source: USDA, World Agricultural Supply and Demand Estimates (WASDE), reports for May 10, 2018, and September 12, 2018.

Notes:

[a] Midpoint of USDA price forecast; marketing year 2018/2019 for crops and calendar 2019 for hogs and milk.

APPENDIX B. TRADE LOSS VERSUS OPPORTUNITY COST

USDA has elected to base MFP payments strictly on estimated trade loss. In contrast, several commodity groups have calculated the "lost market value" and view it as a better measure of the economic damage from the retaliatory tariffs (see "Industry Response to MFP Payment Allocation"). These two "loss" measures are described here.

Trade Loss

Trade loss is the value of lost export sales due to a change in foreign demand (*Table A-3*). With respect to retaliatory tariffs, it is the difference in U.S. agricultural exports with and without the tariffs. It also appears in USDA export forecasts. For example, in May 2018, USDA forecast U.S. agricultural export sales to China for FY2018 of $21.6 billion; by August 2018, USDA had revised its forecast down to $19 billion and initially projected agricultural export sales to China in FY2019 of only $12 billion. Thus, from May to August the U.S. agricultural export outlook to China had declined by $2 billion, while the FY2019 forecast had fallen by as much as $9 billion.

Lost Market Value (or the Opportunity Cost of Missed Sales)

Lost market value describes the opportunity cost of missed sales associated with a drop in market prices. For example, if soybean prices were $10.00 per bushel in March and $8.00 per bushel in October, the opportunity cost of not selling in March (whether from on-farm stocks or by forward contracting the crop in the field) but instead waiting to sell after harvest in October would be $2.00 per bushel. All physical quantities of a commodity available on the farm—including commodities in storage as well as in the field—are potentially subject to a missed sales opportunity. Furthermore, until the producer actually sells the commodity, the realized market value and true opportunity cost remain unknown.

What Is the Correct Cost?

If a trade dispute contributes to a drop in the market price of a commodity, then the associated "lost market value" would affect all quantities of the affected commodity, whether exported or used domestically. This appears to be the type of "loss" being measured by most U.S. commodity groups. However, the retaliatory tariffs are only one of a number of factors that influence market prices. In particular, the outlook for record U.S. soybean and near-record corn harvests in 2018 has likely had an important effect on pressuring market prices lower during the May to September period. This production effect should be excluded from any estimate of trade-based market loss.

Changes in USDA's monthly price forecasts from May to September may provide an upper-bound estimate of the trade impacts (*Table A-4*), since this period coincides with the escalating trade conflicts when the retaliatory tariffs were applied. However, they include the production effect and thus likely overstate any trade impact. According to USDA, during the May-September period, farm prices for MFP commodities declined 18% for soybeans, 8% for sorghum, and 8% for corn but rose 2% for wheat and 15% for cotton. At first glance, these price changes seem out of sync with the MFP payment rates. Sorghum could receive a payment rate that is nearly three times as large as its estimated price decline from May to September. In contrast, corn—which has experienced a price decline identical to sorghum—could receive a payment rate that amounts to 3% of the price decline that corn prices experienced over this same period.[52]

However, given the number of factors influencing market prices over this period, it may not be possible to establish with confidence what market prices would have been in the absence of the retaliatory tariffs. Any viable estimate would have to be generated from a global economic model featuring all major agricultural commodities that compete for land and other inputs in production; may substitute for each other in alternative uses; and captures the interactions of all relevant market factors such as policy, technology, and expected prices, production, and demand. For example, wheat and cotton are to receive per-unit MFP payment rates while experiencing an increase in farm prices during the May-September period. However, 2018 has been a year of poor international wheat harvests,[53] and it could be that wheat prices might have moved to much higher levels in the absence of retaliatory tariffs.

---

[52] G. Schnitkey et al., "Market Facilitation Program: Impacts and Initial Analysis," *farmdoc daily*, August 28, 2018.

[53] Stephanie Bryant-Erdmann, "Weather Continues to Affect Global Wheat Production," U.S. Wheat Associates, September 19, 2018, https://www.uswheat.org/wheatletter/exportable-wheat-supplies-in-the-world-expected-to-fall-forsecond-year-in-a-row/.

In: Agricultural Issues
Editor: Michael Butlin

ISBN: 978-1-53615-474-0

Chapter 4

Farm Programs: Information on Payments*

U.S. Government Accountability Office

441 G St. N.W.
Washington, DC 20548

May 18, 2018

The Honorable Charles E. Grassley Chairman
Committee on the Judiciary
United States Senate

Dear Mr. Chairman:

For each crop year,[1] the U.S. Department of Agriculture (USDA) makes billions of dollars in payments to agricultural producers for which

---

* This is an edited, reformatted and augmented version of U.S. Government Accountability Office, Publication No. GAO-18-384R, dated May 18, 2018.

[1] For these farm programs, USDA makes payments by program, or crop, year—that is, the year in which the crop is harvested. According to USDA, a crop year is the 12-month period starting with the month when the harvest of a specific crop typically begins.

being actively engaged in farming is a requirement.[2] The largest programs in terms of payments are the Price Loss Coverage program, which makes payments in years in which a crop's market price is less than a statutorily set price, and the Agriculture Risk Coverage program, which makes payments in years in which a crop's revenue is less than a revenue guarantee.[3] These payments are to go to producers, both individuals and entities, such as partnerships, corporations, and trusts.[4] Under the Agricultural Act of 2014 (2014 Farm Bill), each member of a farming operation that is a general partnership can generally receive directly or indirectly up to $125,000 per year through the applicable programs if the member meets eligibility requirements, including being determined to be actively engaged in farming.[5]

After publicized instances of farm payments going to individuals not involved in farming, Congress enacted the Agricultural Reconciliation Act of 1987, commonly referred to as the Farm Program Payments Integrity Act.[6] Among other things, the act established eligibility criteria that limit receipt of certain farm program payments to individuals and entities that are "actively engaged in farming." It also established criteria that

---

[2] Being actively engaged in farming is not required for numerous USDA program benefits. For example, being actively engaged in farming is not required for payments made through USDA's conservation and disaster assistance programs or for participation in USDA's federal crop insurance program.

[3] Under the Agricultural Act of 2014 (2014 Farm Bill), Pub. L. No. 113-79, 128 Stat. 649, for each eligible crop, farmers had a one-time choice between participating in the Price Loss Coverage or the Agriculture Risk Coverage programs. Price Loss Coverage payments make up the difference between a crop's statutory "reference price" and its national season average farm price. Agriculture Risk Coverage payments generally make up the difference between a county revenue guarantee and actual crop revenue.

[4] Entities also include other legal organizations such as joint ventures, limited liability companies, limited partnerships, limited liability partnerships, estates, and charitable organizations.

[5] A farming operation is a business enterprise engaged in the production of agricultural products, commodities, or livestock that is operated by an individual legal entity or joint operation. A farming operation member can be either an individual or an entity. According to an FSA handbook, a general partnership is composed of two or more individuals or entities, formed under state law and subject to the terms of a formalized agreement. In a general partnership, members share responsibility for management, profits, and the liability for debts.

[6] The Agricultural Reconciliation Act of 1987 was enacted as title I of the Omnibus Budget Reconciliation Act of 1987, Pub. L. No. 100-203, 101 Stat. 1330. The relevant provisions of the act became effective in the 1989 crop year.

discourage farming operations from avoiding program payment limits by adding individuals or entities to their operation.[7]

For an individual to meet the criteria for being actively engaged in farming under the act, he or she must make significant contributions to a farming operation in two areas: (1) capital, land, or equipment (or some combination of the three) and (2) active personal management or personal labor (or a combination of the two).[8] For an entity, such as a corporation, to meet these criteria, it must separately make a significant contribution of capital, land, or equipment, and its members must collectively make a significant contribution of active personal management or personal labor to the farming operation. For both individuals and entities, their share of a farming operation's profits or losses must be commensurate with their contributions to the farming operation, and those contributions must be at risk.[9] USDA's Farm Service Agency (FSA) is responsible for administering the Farm Program Payments Integrity Act and ensuring that (1) farming operation members meet the actively engaged in farming criteria and (2) they do not receive payments in excess of program payment limits.

In September 2013, we reported that USDA's broad definition of active personal management made it difficult for USDA to determine whether an individual's contribution is significant.[10] We also reported that, under this broad definition, management responsibilities could be distributed among farming operation members to increase the number of members who could claim eligibility for payments. For example, we

---

[7] Congress approved statutory limits for certain farm program payments that established, for various program payment types, the maximum amount that an individual or entity can receive each year.

[8] USDA regulations define active personal management to include such tasks as arranging financing for the operation, supervising the planting and harvesting of crops, and marketing the crops. A landowner may be considered actively engaged in farming with respect to the owned land without providing a significant contribution of personal labor or active personal management, so long as he or she received rent or income based on the land's production or the farming operation's results, his or her share of the profits is commensurate with the landowner's contributions, and those contributions are at risk.

[9] For such contributions to be considered at risk, there must be a possibility that the individual or member of an entity could suffer a financial loss.

[10] GAO, *Farm Programs: Changes Are Needed to Eligibility Requirements for Being Actively Involved in Farming*, GAO-13-781 (Washington, D.C.: Sept. 26, 2013).

reported that a farming operation received payments of about $651,000 in 2012. This farming operation had 22 members, and 16 of these members claimed contributions of active personal management. We recommended that Congress consider modifying the definition of significant contributions of management activities.

The 2014 Farm Bill required USDA to promulgate new regulations to define a significant contribution of active personal management for farming operations that were not comprised entirely of family members (nonfamily farming operations). The 2014 Farm Bill also directed that farming operations comprised solely of family members be exempt from the new restrictions on contributions of active personal management in the new regulations. In 2015, USDA issued new regulations.[11] Among other things, these regulations added a new, more specific definition for "active personal management." This definition includes a list of critical management activities that qualify as a significant contribution if such activities are annually performed to either of the minimum levels established (500 hours or 25 percent of the total management hours required for the operation).

For nonfamily farming operations, the 2014 Farm Bill directed USDA to determine whether it would be appropriate to establish limits on the number of persons per farming operation who may be considered actively engaged in farming based on a significant contribution of active personal management. USDA's subsequent 2015 regulations limit nonfamily farming operations—specifically general partnerships and joint ventures—to no more than three members qualifying for payments by contributing active personal management or a combination of management and labor.[12] Farming operations comprised entirely of family members are exempt from the new restrictions on contributions of active personal management

---

[11] Payment Limitation and Payment Eligibility; Actively Engaged in Farming, 80 Fed. Reg. 78119 (Dec. 16, 2015).

[12] FSA defines a joint venture as a short-term association of individuals or entities, where the association exists without an actual partnership. The joint venture's members share the liability for debts.

and have no limit on the potential number of members qualifying for payments.[13]

You asked us to update selected information from our September 2013 report, specifically information on farm program payments to farming operations and the contributions that these operations' members claimed in order to qualify as being actively engaged in farming. This chapter responds, in part, to your request.[14] Our objective was to provide 2015 crop year information on (1) the distribution and amount of payments to farming operations by type of entity, (2) entities with the highest farm program payments, and (3) general partnership members' claimed contributions of active personal management or personal labor.

To address this objective, we obtained and analyzed FSA data on applicable farm program payments made to entities for the 2015 crop year, which was the most recent year for which data were available at the time we obtained data for our analysis. In addition, we analyzed FSA data on farming operation members' contributions of active personal management, personal labor, or a combination of both to farming operations to meet the requirements to be actively engaged in farming and FSA data on applicable farm program payments attributed to these members. We also analyzed the data to determine the payments and contributions claimed in farming operations, categorized by the number of farming operation members. To assess the reliability of both FSA datasets, we reviewed our previous assessments of the data's reliability. In addition, we (1) performed electronic tests of pertinent data elements, (2) reviewed information about the data, and (3) interviewed agency officials knowledgeable about the data. We determined that the data were sufficiently reliable for the purposes of determining the distribution and amount of payments to

---

[13] The 2015 regulations apply to eligibility for payments earned for the 2016 crop year for farming operations with only 2016 spring-planted crops and to eligibility for payments for the 2017 and subsequent crop years for all farming operations (those with either spring- or fall-planted crops). USDA estimated that the new regulations on nonfamily general partnerships and joint ventures would affect about 3,200 operations for the 2016 through 2018 crop years.

[14] We have additional ongoing work on requirements for being actively engaged in farming that responds to the remainder of the request.

farming operations based on members' claims of active personal management or personal labor contributions.

To obtain information about the number of members who claimed contributions of management or labor to the 20 farming operations that received the highest payments for 2015,[15] we contacted the seven FSA state offices responsible for administering payment eligibility requirements for these farming operations. We also obtained information about these operations for 2016.[16] To assess the reliability of the information that we obtained from FSA state offices, we compared it with information from the FSA datasets. We determined that the data were sufficiently reliable for our purpose.

We conducted this performance audit from July 2017 to May 2018 in accordance with generally accepted government auditing standards. Those standards require that we plan and perform the audit to obtain sufficient, appropriate evidence to provide a reasonable basis for our findings and conclusions based on our audit objectives. We believe that the evidence obtained provides a reasonable basis for our findings and conclusions based on our audit objectives.

BACKGROUND

To be considered actively engaged in farming, the Farm Program Payments Integrity Act requires that an individual or entity provide the farming operation with a significant contribution of inputs of capital, land, or equipment as well as a significant contribution of services of active personal management or personal labor. Figure 1 shows the input and

---

[15] We analyzed FSA data to identify the 20 farming operations that received the highest payments for 2015. We did not obtain information for one of these farming operations because the applicable FSA state office was experiencing an unusually heavy workload at the time we contacted the state offices in February 2018.

[16] Although the 2015 crop year was the most recent year for which data were available at the time we obtained data from FSA's data center for our analysis, FSA state offices had data for the 2016 crop year at the time we contacted them in February 2018.

service contribution requirements that farming operation members must meet to be considered actively engaged in farming.

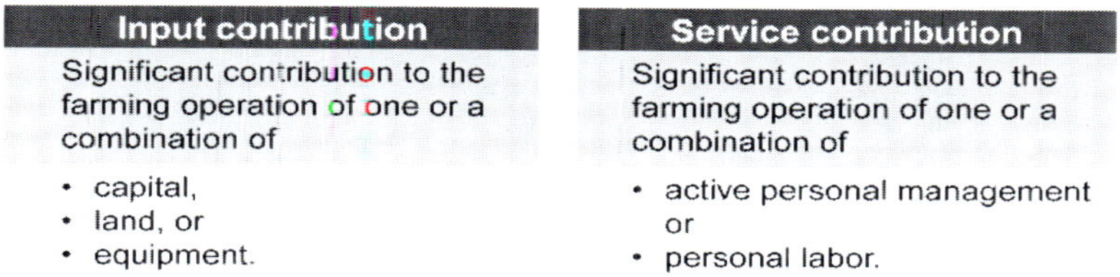

Source: GAO analysis of the Farm Program Payments Integrity Act and the Farm Service Agency's habdbook on payment eligibility and payment limits. | GAO-18-384R.

Figure 1. Input and Service Contribution Requirements for Farming Operation Members to Be Considered Actively Engaged in Farming.

As stated in FSA's handbook on payment eligibility and payment limits, to satisfy the actively engaged in farming criteria, an active personal management contribution must, among other things, be critical to the profitability of the farming operation.[17] For personal labor, the contribution is to be an amount that is the smaller of 1,000 hours annually or 50 percent of the total hours that would be required to conduct a farming operation comparable in size to the individual's or legal entity's commensurate share in the farming operation.

In addition to meeting the input and service contribution requirements, under current regulations, a farming operation member's contributions to the farming operation must be in proportion to his or her share of the operation's profits and losses, and the contributions must be "at risk."[18] For a member's contribution to be considered at risk, there must be a possibility that the member could suffer a financial loss. For example, if a member of a general partnership or joint venture receives a guaranteed payment for any part of a contribution of labor or management, that

---

[17] U.S. Department of Agriculture, Farm Service Agency, *Payment Eligibility, Payment Limitation, and Average Adjusted Gross Income—Agricultural Act of 2014*, 5-PL, Amendment 3.

[18] 7 C.F.R. §§ 1400.201(b)(1); 1400.204(a).

contribution is not at risk and is to be excluded in determining whether that member is actively engaged in farming.

In addition to the requirements specified in the act, FSA's handbook and its website provide direction for actively engaged in farming determinations involving spouses. According to FSA, both spouses may be considered "actively engaged in farming" and qualify for farm program payments if one spouse makes the requisite contributions to meet the actively engaged in farming requirements.[19]

Table 1. Farm Programs and Types of Payments for Which Being Actively Engaged in Farming Applies and Payment Limits for These Programs and Payment Types

Farm programs/payment types	Payment limit per crop year (in dollars)
Price Loss Coverage, Agriculture Risk Coverage, loan deficiency payments, marketing loan gains[a]	
• for crops other than peanuts	125,000
• for peanuts	125,000
Cotton Transition Assistance Program[b]	40,000
Cotton Ginning Cost-Share Program[c]	40,000

Source: GAO analysis of Farm Service Agency documents. | GAO-18-384R.

Notes: The payment limit for general partnerships and joint ventures is per member or partner who meets payment eligibility requirements, including being actively engaged in farming. The payment limit for corporations, limited liability companies, and other entity types is per entity that meets payment eligibility requirements. An eligible individual or entity can receive payments up to the limit from one or more of the programs for which being actively engaged in farming applies.

[a]A farming operation realizes a marketing loan gain if it repays a marketing assistance loan at less than the amount of the loan principal. Similarly, a farming operation can realize a gain through the use of a commodity exchange certificate. Being actively engaged in farming and payment limits do not apply to a gain realized from the use of commodity exchange certificates.

[b]Cotton Transition Assistance Program payments were available for the 2014 crop year and, for certain counties, the 2015 crop year.

[c]Cotton Ginning Cost-Share Program payments were available for the 2015 and 2016 crop years.

Table 1 shows the farm programs and types of payments for which the requirement of being actively engaged in farming applies and the payment

[19] U.S. Department of Agriculture, Farm Service Agency, *Actively Engaged in Farming*, accessed April 11, 2018, https://www.fsa.usda.gov/programs-and-services/payment-eligibility/actively_engaged/index.

limits for these programs based on our analysis of FSA documents. An eligible individual or entity can receive payments from one or more of these programs. Each payment limit shown in the table is a separate payment limit. For example, an individual could receive $125,000 for crops other than peanuts and an additional $125,000 for peanuts.

INFORMATION ON FARM PROGRAM PAYMENTS TO FARMING OPERATIONS AND CONTRIBUTIONS FARMING OPERATION MEMBERS CLAIMED TO QUALIFY AS BEING ACTIVELY ENGAGED IN FARMING

For those USDA program payments requiring active engagement in farming, we determined that

- USDA distributed about $2.7 billion in payments to 95,417 entities, such as corporations, general partnerships, joint ventures, and limited liability companies;
- USDA distributed an average of $884,495 in payments to the 50 farming operations receiving the highest payments for 2015; and
- general partnership members' payments were predominantly based on members' claimed contributions of combined management and labor (74.6 percent) and management (23.1 percent), while labor was 2.3 percent.

Distribution and Amount of Farm Program Payments by Type of Entity

For 2015, USDA's farm programs distributed approximately $2.7 billion in payments subject to actively engaged in farming requirements to 95,417 entities. Table 2 categorizes these entities and shows the distribution of payments made by entity type for 2015. General partnerships received the highest total and average payments. Specifically, 22,636 general partnerships received about $1.3 billion, or nearly half of the total payments made to entities, with an average payment of $56,760 per entity. Joint ventures received average payments of $43,282. Corporations received average payments of $25,928, and limited liability companies received average payments of $15,532.

Table 2. Distribution and Amount of Farm Program Payments by Type of Entity, for 2015

Type of entity	Entities		Payments (dollars)		
	Number	Percentage	Total	Percentage	Average per entity
General partnerships	22,636	23.7	1,284,815,722	46.9	56,760
Joint ventures	4,783	5.0	207,016,584	7.6	43,282
Corporations	30,335	31.8	786,531,333	28.7	25,928
Limited liability companies	23,216	24.3	360,585,017	13.2	15,532
Other[a]	14,447	15.1	102,576,853	3.7	7,100
Total	95,417	100.0[b]	2,741,525,508	100.0[b]	28,732

Source: GAO analysis of Farm Service Agency data. | GAO-18-384R.

Notes: Farm program payments to sole proprietorships are not included in this table. When these payments are included, the total payments that were subject to actively engaged in farming requirements for 2015 were about $8.6 billion.

The farm program payments in this table are payments that are subject to actively engaged in farming requirements. However, we did not include marketing loan gains—gains that a farming operation realizes if it repays a marketing assistance loan at less than the amount of the loan principal—because the data we received from the Farm Service Agency did not include them. For 2015, nationwide marketing loan gains were about 2 percent of the total payments ($8.6 billion) that were subject to actively engaged in farming requirements.

[a]Includes limited partnerships, irrevocable and revocable trusts, estates, churches, charities, and nonprofit organizations.

[b]The percentages may not total to 100 because of rounding.

Payments and Member Contributions for General Partnerships and Joint Ventures with Highest Farm Program Payments

USDA's farm programs distributed an average of $884,495 in payments to the 50 farming operations receiving the highest payments for 2015, 49 of which are general partnerships; 1 is a joint venture. Table 3 categorizes the 50 farming operations into groups of 10 ranked by payment. As shown in the table, the payments were primarily based on individuals claiming contributions of active personal management or a combination of active personal management and personal labor. (See encl. for additional information on 19 of 20 general partnerships and joint ventures that received the highest farm program payments for 2015.)

Table 3. Payments and Claimed Contributions for General Partnerships and Joint Ventures with Highest Farm Program Payments, by Ranked Group of 10, for 2015

Payment rank	Average payment per general partnership or joint venture (dollars)	Average number of members[a]	Average number of members claiming management or labor contributions, by type[b]		
			Active personal management only	Personal labor only	Combination of active personal management and personal labor
1-10	1,339,065	13.6	8.0	0.1	2.0
11-20	862,195	7.6	5.3	0.0	1.4
21-30	796,746	8.0	2.0	0.0	4.5
31-40	732,037	6.3	2.1	0.6	2.5
41-50	692,433	6.0	1.9	0.0	3.1
1-50	884,495	8.3	3.9	0.1	2.7

Source: GAO analysis of Farm Service Agency data. | GAO-18-384R.

Notes: Among the 50 farming operations, 49 are general partnerships and 1 is a joint venture. The farm program payments in this table are payments that are subject to actively engaged in farming requirements. However, we did not include marketing loan gains—gains that a farming operation realizes if it repays a marketing assistance loan at less than the amount of the loan principal—because the data we received from the Farm Service Agency did not include them. For 2015, nationwide marketing loan gains were about 2 percent of the total payments that were subject to actively engaged in farming requirements.

[a]The number of members consists of the number of entities and individuals who are members of the general partnership or joint venture. The number of members is not equal to the sum of active personal management only, personal labor only, or combination of the two because the number of members includes (1) entities—entities cannot provide management or labor—and (2) individuals who did not claim contributions of personal management or personal labor. For example, if spouses are farming together in a general partnership and one spouse is determined to be making a significant contribution of active personal management, the other spouse is credited with a significant contribution of active personal management.

[b]Members can be individuals or entities. However, only members who are individuals can make contributions of personal management or personal labor.

Table 4. Farm Program Payments and Contributions Claimed in General Partnerships, by Number of Members, for 2015

Number of members	Number of general partnerships	Payments (dollars)	Type of contribution					
			Active personal management only		Personal labor only		Combination of active personal management and personal labor	
			Payments (dollars)	%	Payments (dollars)	%	Payments (dollars)	%
1	486	7,504,982	2,558,034	34.1	125,648	1.7	4,821,299	64.2
2	12,213	501,134,456	67,183,912	13.4	7,059,889	1.4	426,890,655	85.2
3-5	8,287	493,937,208	125,202,189	25.3	14,566,907	2.9	354,168,111	71.7
6-10	1,107	111,769,863	54,614,878	48.9	4,349,072	3.9	52,805,913	47.2
11 or more	150	12,086,225	10,259,152	84.9	170,435	1.4	1,656,637	13.7
Total	22,243[a]	1,126,432,734[a]	259,818,165	23.1	26,271,952	2.3	840,342,616	74.6

Legend: % = Percentage of payments for that type of contribution.

Source: GAO analysis of Farm Service Agency data. | GAO-18-384R.

Notes: The farm program payments in this table are subject to actively engaged in farming requirements. However, we did not include marketing loan gains—gains that a farming operation realizes if it repays a marketing assistance loan at less than the amount of the loan principal—because the data we received from the Farm Service Agency did not include them. For 2015, nationwide marketing loan gains were about 2 percent of the total payments that were subject to actively engaged in farming requirements.

The percentages of payments based on contributions in this table may not total to 100 because of rounding.

[a]Not all general partnerships and their related contributions data are included in the Farm Service Agency's contributions database. For some of the general partnerships included in the Farm Service Agency's contributions database, no contribution was identified. We did not include these general partnerships and their payments in this table.

Payments and Contributions Claimed in General Partnerships by Number of Members

For 2015, general partnerships received a high proportion of their payments based on members' claims of contributing active personal management, either management only (23.1 percent) or management in combination with labor (74.6 percent). Table 4 shows the payments and contributions claimed in general partnerships by the number of members and type of contribution.

Enclosure

Table 5 lists additional information on 19 of 20 general partnerships and joint ventures that received the highest payments for the 2015 crop year.[20] For each of these farming operations, the table shows the Census Bureau region in which it is located; the operation's total payments that are subject to being actively engaged in farming; the three crops for which the operation received the highest payments (ordered by payment amount); the number of members (individuals and entities); and the number of members (individuals) claiming contributions of active personal management, personal labor, or a combination of both.

Table 6 lists the general partnerships and joint ventures in the same order as table 5, but provides 2016 data about these operations. As previously mentioned, the U.S. Department of Agriculture's (USDA) 2015 regulations limit nonfamily farming operations to no more than three members qualifying for payments by contributing active personal management or a combination of such management and personal labor. Farming operations comprised entirely of family members are exempt from the new restrictions on contributions of active personal management.

[20] We analyzed Farm Service Agency (FSA) data to identify the 20 general partnerships and joint ventures that received the highest payments for 2015. As previously stated, we did not obtain information for one of these farming operations because the applicable FSA state office was experiencing an unusually heavy workload at the time we contacted the state offices in February 2018.

Table 5. Payments and Claimed Contributions of Management and Labor for Selected Examples of General Partnerships and Joint Ventures for 2015

					Number of members claiming management or labor contributions, by type[c]		
Example	Census Bureau region[a]	Payments (dollars)	Primary crops	Number of members[b]	Active personal management only	Personal labor only	Combination of active personal management and personal labor
1	Midwest and South	3,708,108	Rice, soybean, corn	34 (32 corporations and 2 individuals)	25 (plus 10 spouses)	0	0
2	South	1,425,965	Rice, soybean, corn	11 (all LLCs)	10 (plus 2 spouses)	0	0
3	Midwest	1,259,253	Corn, soybean, wheat	11 (7 corporations and 4 individuals)	9 (plus 2 spouses)	0	0
4	South	1,242,019	Peanuts, soybean, canola	8 (all LLCs)	0	1	5 (plus 2spouses)
5	South	1,241,497	Peanuts, corn, sorghum	7 (all individuals)	6	1	0
6	South	1,100,910	Soybean, rice, corn	15 (9 corporations and 6 individuals)	10 (plus 5 spouses)	0	0
7	South	1,085,836	Corn, sorghum, wheat	8 (all individuals)	0	0	5 (plus 3spouses)

					Number of members claiming management or labor contributions, by type[c]		
Example	Census Bureau region[a]	Payments (dollars)	Primary crops	Number of members[b]	Active personal management only	Personal labor only	Combination of active personal management and personal labor
8	Midwest	1,052,305	Rice, corn, wheat	12 (10 corporations and 2 individuals)	10 (plus 2 spouses)	0	0
9	South	937,700	Corn, soybean, cotton	22 (all LLCs)	18 (plus 4 spouses)	0	0
10	South	934,739	Rice, soybean, sorghum	10 (7 corporations and 3 LLCs)	8 (plus 3 spouses)	0	0
11	South	923,208	Rice, soybean, wheat	8 (all individuals)	4 (plus 4 spouses)	0	0
12	South	902,808	Peanuts, cotton, corn	4 (all LLCs)	0	0	4
13	South	896,065	Peanuts, corn, wheat	5 (all LLCs)	2	0	3
14	South	891,316	Rice, sorghum, corn	9 (all corporations)	15 (plus 4 spouses)	0	0
15	Midwest	870,954	Corn, soybean, wheat	8 (all LLCs)	4 (plus 4 spouses)	0	0
16	South	842,437	Rice, corn, soybean	9 (all corporations)	5 (plus 4 spouses)	0	0
17	South	827,977	Corn, wheat, sorghum	10 (all individuals)	0	0	8 (plus 2spouses)

Table 5. (Continued)

					Number of members claiming management or labor contributions, by type[c]		
Example	Census Bureau region[a]	Payments (dollars)	Primary crops	Number of members[b]	Active personal management only	Personal labor only	Combination of active personal management and personal labor
18	Midwest	827,477	Corn, soybean, wheat	10 (8 corporations and 2 individuals)	0	1 (plus 1 spouse)	4 (plus 4spouses)
19	South	791,731	Peanuts, cotton, corn	5 (all individuals)	0	0	5

Legend: LLC = limited liability company.

Sources: Farm Service Agency state office officials and GAO analysis of Farm Service Agency data. | GAO-18-384R.

Notes: We analyzed Farm Service Agency data to identify the 20 general partnerships and joint ventures that received the highest payments for 2015. We did not obtain information for one of these farming operations because the applicable Farm Service Agency state office was experiencing an unusually heavy workload at the time we contacted the state offices in February 2018. These payment amounts provided by officials in Farm Service Agency state offices include marketing loan gains—gains that a farming operation realizes if it repays a marketing assistance loan at less than the amount of the loan principal.

Members of a general partnership or joint venture can be individuals or entities (e.g., corporations or limited liability companies). Each member represents one limitation for payment limitation purposes.

Both spouses may be considered "actively engaged in farming" and qualify for farm program payments if one spouse makes the requisite contributions to meet the actively engaged in farming requirements.

[a]The U.S. Census Bureau divides the 50 states among 4 regions—Northeast, South, Midwest, and West. The Northeast region includes Connecticut, Maine, Massachusetts, New Hampshire, New Jersey, New York, Pennsylvania, Rhode Island, and Vermont. The South region includes Alabama, Arkansas, Delaware, District of Columbia, Florida, Georgia, Kentucky, Louisiana, Maryland, Mississippi, North Carolina, Oklahoma, South Carolina, Tennessee, Texas, Virginia, and West Virginia. The Midwest region includes Indiana, Illinois, Iowa, Kansas, Michigan, Minnesota, Missouri, Nebraska, North Dakota, Ohio, South Dakota, and Wisconsin. The West region includes Alaska, Arizona, California, Colorado, Hawaii, Idaho, Nevada, New Mexico, Montana, Oregon, Utah, Washington, and Wyoming.

[b]The number of members consists of the number of entities and individuals who are members of the general partnership or joint venture.

[c]Individuals claim contributions of active personal management and personal labor and can provide these contributions on behalf of entities within a general partnership or joint venture.

Table 6. Payments and Claimed Contributions of Management and Labor for Selected Examples of General Partnerships and Joint Ventures for 2016

Example	Census Bureau region[a]	Payments (dollars)	Number of members[b]	Number of members claiming management or labor contributions, by type[c]		
				Active personal management only	Personal labor only	Combination of active personal management and personal labor
1	Midwest and South	2,190,115	8 (6 corporations and 2 individuals)	4 (plus 4 spouses)	0	0
2	South	1,132,310	11 (all LLCs)	10 (plus 2 spouses)	0	0
3	Midwest	344,692	11 (7 corporations and 4 individuals)	9 (plus 2 spouses)	0	0
4	South	993,216	9 (all LLCs)	0	0	5 (plus 4 spouses)
5	South	1,213,094	9 (all individuals)	8	0	1
6	South	466,005	6 (all individuals)	3 (plus 3 spouses)	0	0
7	South	873,978	8 (all individuals)	0	0	5 (plus 3 spouses)
8	Midwest	1,389,537	12 (10 corporations and 2 individuals)	0	0	10 (plus 2 spouses)
9	South	149,839	5 (all LLCs)	3 (plus 3 spouses)	0	0

Table 6. (Continued)

Example	Census Bureau region[a]	Payments (dollars)	Number of members[b]	Number of members claiming management or labor contributions, by type[c]		
				Active personal management only	Personal labor only	Combination of active personal management and personal labor
10	South	576,677	5 (3 LLCs and 2 corporations)	4 (plus 1 spouse)	0	0
11	South	931,924	8 (all individuals)	4 (plus 4 spouses)	0	0
12	South	690,682	7 (all LLCs)	0	0	4 (plus 3 spouses)
13	South	735,189	5 (all LLCs)	1	0	4
14	South	998,811	9 (all corporations)	11 (plus 4 spouses)	0	0
15	Midwest	608,288	6 (all LLCs)	3 (plus 3 spouses)d	0	0
16	South	1,048,498	9 (all corporations)	5 (plus 4 spouses)	0	0
17	South	1,082,576	10 (all individuals)	0	0	8 (plus 2 spouses)
18	Midwest	552,262	10 (8 corporations and 2 individuals)	0	1 (plus 1 spouse)	4 (plus 4 spouses)
19	South	506,258	4 (all individuals)	0	0	4

Legend: LLC = limited liability company.

Source: Farm Service Agency state office officials | GAO-18-384R.

Notes: All of these examples are family operations, except examples 3, 8, 16, and 18, which are nonfamily operations.

Members of a general partnership or joint venture can be individuals or entities (e.g., corporations or limited liability companies). Each member represents one limitation for payment limitation purposes.

Although the 2015 crop year was the most recent year for which data were available at the time we obtained data for our analysis, Farm Service Agency state offices had data for the 2016 crop year at the time we contacted them in February 2018.

We did not include the three leading crops in 2016 payments for these examples because the information was not readily available.

Both spouses may be considered "actively engaged in farming" and qualified for farm program payments if one spouse makes the requisite contributions to meet the actively engaged in farming requirements.

[a]The U.S. Census Bureau divides the 50 states among 4 regions—Northeast, South, Midwest, and West. The Northeast region includes Connecticut, Maine, Massachusetts, New Hampshire, New Jersey, New York, Pennsylvania, Rhode Island, and Vermont. The South region includes Alabama, Arkansas, Delaware, District of Columbia, Florida, Georgia, Kentucky, Louisiana, Maryland, Mississippi, North Carolina, Oklahoma, South Carolina, Tennessee, Texas, Virginia, and West Virginia. The Midwest region includes Indiana, Illinois, Iowa, Kansas, Michigan, Minnesota, Missouri, Nebraska, North Dakota, Ohio, South Dakota, and Wisconsin. The West region includes Alaska, Arizona, California, Colorado, Hawaii, Idaho, Nevada, New Mexico, Montana, Oregon, Utah, Washington, and Wyoming.

[b]The number of members consists of the number of entities and individuals who are members of the general partnership or joint venture.

[c]Individuals claim contributions of active personal management and personal labor and can provide these contributions on behalf of entities within a general partnership or joint venture.

USDA's 2015 regulations apply to eligibility for payments earned for the 2016 crop year for farming operations with only 2016 spring-planted crops. The regulations also apply to payments for the 2017 and subsequent crop years for all farming operations (i.e., those with either spring- or fall-planted crops). Farm Service Agency (FSA) officials said that because the 2015 regulations were fully implemented for the 2017 crop year, their effects on payments to farming operations are likely to be more apparent after USDA issues payments for the 2017 crop year and when data on those payments becomes available.[21] In table 6, examples 3, 8, 16, and 18 are nonfamily farming operations, and the other examples are family farming operations.

[21] According to FSA officials, USDA will complete the issuance of most 2017 crop year payments late in calendar year 2018.

In: Agricultural Issues
Editor: Michael Butlin

ISBN: 978-1-53615-474-0

Chapter 5

SEED COTTON AS A FARM PROGRAM CROP: IN BRIEF*

Randy Schnepf

SUMMARY

The Bipartisan Budget Act of 2018 (P.L. 115-123; BBA), signed into law on February 9, 2018, included a provision—Section 60101(a)—which amended the 2014 farm bill (P.L. 113-79) to add seed cotton as a "covered commodity," thus making cotton eligible for the Price Loss Coverage (PLC) and Agricultural Risk Coverage (ARC) farm revenue support programs.

The 2014 farm bill provides authority for farm programs for the 2014 through 2018 crop years. Under the 2014 farm bill as signed into law in February 2014, neither cotton nor its co-product, cottonseed, were eligible to participate in the newly created ARC and PLC programs. During the first three years of the 2014 farm bill (crop years 2014 through 2016), nearly $20 billion in support payments have been made to eligible program crops under ARC and PLC.

* This is an edited, reformatted and augmented version of Congressional Research Service Publication No. R45143, Updated August 1, 2018.

Upland cotton had been eligible for U.S. Department of Agriculture (USDA) farm support programs since their origin in the 1930s. However, in response to a trade dispute with Brazil, the 2014 farm bill permanently removed upland cotton from eligibility for the ARC and PLC programs. Instead, the remaining cotton-specific support programs were supplemented by offering the U.S. cotton sector new transitional payments and, in 2015, a new shallow-loss revenue insurance product: the Stacked Income Protection Program (STAX).

Since early 2014, U.S. cotton producers have been largely unsatisfied with their economic situation due to falling cotton prices and the perceived lack of revenue support offered under STAX. In early 2015, the U.S. cotton sector began advocating to be made eligible for payments under the ARC and PLC programs. Several Members of the House Agriculture Committee (including both its chairman and ranking member) were supportive of this effort and actively sought a legislative opportunity to reintegrate either cottonseed or seed cotton—the harvested but un-ginned cotton boll that includes both lint and cottonseed—back into the revenue support programs in the 2014 farm bill. After several legislative attempts, this effort was successful when Congress passed the BBA with a provision that amended the 2014 farm bill to add seed cotton as a "covered commodity," thus making it eligible for either ARC or PLC beginning with the 2018 crop.

To participate in the seed cotton program, producers must first make three decisions specified in the new statute: what portion of their farm program generic base acres will be designated as cotton-specific base, whether to update historical cotton program yields (needed to determine the per-acre payment rate), and which program to participate in—ARC or PLC.

The Congressional Budget Office (CBO) projects the cost of adding seed cotton as a covered commodity at nearly $3 billion over 10 years. However, accompanying budget offsets are projected to lower the net cost to $61 million.

The policy change for cotton has potential implications for both domestic and international commodity markets. However, perhaps the most critical aspect of the new seed cotton policy is that, as part of current law, it becomes part of CBO's budget baseline used to write the next farm bill. As a result, it avoids the potential policy debate surrounding a search for budgetary offsets that would otherwise be needed to pay for the costs associated with adding a new farm program. This result could facilitate reaching an agreement among congressional negotiators working on the farm support programs in Title I of a new farm bill.

BACKGROUND

This chapter briefly describes cotton's special treatment, relative to other traditional farm program crops, in the 2014 farm bill. It also reviews the legislative attempts—which successfully culminated with the Bipartisan Budget Act of 2018 (P.L. 115-123; BBA)—to reintegrate cotton as a program crop eligible for major farm revenue support programs. Then, it reviews the specifics of the new seed cotton program and discusses three potential issues related to the establishment of seed cotton as a program crop.

The WTO Brazil-U.S. Cotton Trade Dispute

As a member of the World Trade Organization (WTO), the United States has committed to abide by WTO rules and disciplines.[1] WTO rules covering agricultural trade and domestic support programs for agriculture have played a large role in shaping current U.S. cotton policy.[2] In particular, in 2002, Brazil—a major cotton export competitor—initiated a long-running WTO dispute settlement case against U.S. cotton support programs. Brazil charged that these programs were depressing international cotton prices and thus artificially and unfairly reducing the quantity and value of Brazil's cotton exports, causing economic harm to its cotton sector. In 2004, a WTO panel ruled against the United States in the cotton case. As a result of the ruling and the potential for WTO-sanctioned retaliation, the United States made several policy changes—including both legislative action and administrative adjustments—in an attempt to bring the related cotton support programs into WTO compliance. Despite these efforts, Brazil successfully argued that the United States had failed to fully comply with the WTO panel's ruling. As a result, in 2009 the WTO granted Brazil authority to retaliate against U.S. goods and services.[3]

---

[1] See CRS In Focus IF10192, *WTO Disciplines of Domestic Support for Agriculture*.

[2] For more information, see CRS In Focus IF10193, *The WTO Brazil-U.S. Cotton Case*.

[3] See CRS Report R43336, *The WTO Brazil-U.S. Cotton Case*.

With trade retaliation in the offing, the United States removed upland cotton from eligibility for the PLC and ARC revenue support programs in the 2014 farm bill. Then, in October 2014, the two countries signed a Memorandum of Understanding (MOU) that finalized a mutual resolution to the case based on U.S. cotton policy as enacted in the 2014 farm bill.[4] By agreeing to a mutual MOU, the two countries avoided trade retaliation. In particular, according to Section VI of the MOU—the Peace Clause—Brazil would refrain from retaliation prior to September 30, 2018, providing that the United States limits cotton payments to those programs specifically authorized by the 2014 farm bill such as STAX and MAL.

Cotton in the 2014 Farm Bill

Under the 2014 farm bill, upland cotton was made ineligible for new revenue support programs available to traditional program crops—referred to as covered commodities[5]—and was given a reduced marketing loan rate.[6] Instead, upland cotton producers were eligible for new temporary transition payments and a stand-alone, county-based revenue insurance policy called the Stacked Income Protection Plan (STAX).[7] Cotton producers also remained eligible for several smaller (in terms of payments) support programs that were unaffected by the WTO case.

---

[4] WTO, "United States—Subsidies on Upland Cotton," notification of a mutually agreed solution (WT/DS267/46), October 23, 2014, https://www.wto.org/english/tratop_e/dispu_e/cases_e/ds267_e.htm.

[5] Covered commodities refer to those crops eligible to participate in either the PLC or ARC programs. Under the provisions of the 2014 farm bill, 7 U.S.C. 9011(6), the term *covered commodity* means wheat, oats, and barley (including wheat, oats, and barley used for haying and grazing), corn, grain sorghum, long grain rice, medium grain rice, pulse crops, soybeans, other oilseeds, and peanuts.

[6] For details, see CRS In Focus IF10714, *Farm Bill Primer: The Marketing Assistance Loan Program*.

[7] For details, see CRS Report R43494, *Crop Insurance Provisions in the 2014 Farm Bill (P.L. 113-79)*.

Cotton Excluded from New Revenue Programs

The 2014 farm bill established two new support programs—the Agriculture Risk Coverage (ARC) and Price Loss Coverage (PLC) programs—were established.[8] The new revenue programs provide support to U.S. agricultural producers of covered commodities through either statutory reference prices under PLC or historical revenue guarantees based on a five-year moving average of crop prices and yields under ARC. Payments under ARC and PLC are decoupled from current plantings, meaning they are not based on current production. The decoupled payment feature was added in earlier farm bills to comply with WTO rules governing domestic support programs. Instead, payments are made on historical program acres (known as base acres)[9] for a select list of program crops referred to as covered commodities. Cotton was explicitly removed from the list of covered crops, thus making cotton base acres (estimated at 17.6 million acres) ineligible for participation in the ARC and PLC programs.

Historical Cotton Program Acres Renamed as Generic Base

To avoid having cotton producers completely lose access to farm program benefits on their historic cotton base acres, those acres were renamed "generic base"[10] and retained in a producer's total acreage base. As such, they were eligible for program payments under ARC and PLC but only if a covered crop (excluding upland cotton) was planted on the generic base.[11] Thus, ARC and PLC payments made on generic base were coupled to actual plantings of program crops.[12] The "coupling of ARC/PLC payments to specific crops" provided strong incentives to plant other eligible covered crops on generic base acres in lieu of upland cotton.

---

[8] For details, see CRS In Focus IF10711, *Farm Bill Primer: ARC and PLC Support Programs.*

[9] Base acres are historical acres planted to USDA program crops as recorded by USDA's Farm Service Agency (FSA).

[10] 7 U.S.C. 9011(9).

[11] 7 U.S.C. 9011(4)(B).

[12] This distinction is important for another reason: the program outlays made on generic base represent a potential offset when considering how to establish a new cotton program with baseline funding, as discussed later in this report.

Cotton-Specific Programs Remaining in the 2014 Farm Bill

Producers of upland cotton remain eligible for support under several other farm programs under the 2014 farm bill including benefits under the following programs:

- Marketing Assistance Loan (MAL) Program for Cotton Producers. The MAL program provides both a floor price and interim financing for so-called loan commodities. A participating producer may put a harvested "loan" crop under a nine-month, nonrecourse loan valued at a statutory commodity loan rate. The loan uses the crop as collateral, and the loan rate, in effect, establishes a price guarantee. Special benefits are available if the adjusted world price— announced weekly by USDA—falls below the cotton loan rate. The 2014 farm bill changed the loan rate for upland cotton from $0.52 per pound to the simple average of the adjusted prevailing world price for the immediately preceding two market years but not less than $0.45 per pound or more than $0.52 per pound.[13] During the first two years of the 2014 farm bill, cotton producers received $700 million in payments under the MAL program—$372 million in 2014 and $328 million in 2015. More recent data are not yet available.
- CTAP. The 2014 farm bill also established CTAP as an offset for the loss of ARC and PLC program eligibility. Upland cotton producers were eligible for CTAP payments on a percentage of former cotton base acres for crop years 2014 and 2015—60% of base received payments in 2014, 36.5% of base in 2015. CTAP outlays were a combined $484 million during 2014 and 2015. Transition payments ended after the 2015 crop.
- CGCS. In 2016, USDA used its administrative authority under the Commodity Credit Corporation (CCC) Charter Act to offer U.S. upland cotton producers CGCS payments valued at $328 million. The CGCS program provided cost-share payments for cotton

---

[13] For details on the MAL, see CRS In Focus IF10714, *Farm Bill Primer: The Marketing Assistance Loan Program.*

producers' cotton ginning costs based on their 2015 cotton plantings multiplied by 40% of average regional ginning costs.[14] The CGCS program was offered again in 2018 with similar payment terms but at a reduced 20% share of average regional ginning costs based on their 2016 cotton plantings.[15] This second round is projected to result in $227 million in CGCS payments in 2018.[16]

- Economic Adjustment Assistance for Domestic Users (EAAU) Program (P.L. 113-79, §1207(c)). EAAU payments are made to domestic users for all documented use of upland cotton on a monthly basis regardless of the origin of the upland cotton. The payment rate is $0.03 per pound and results in CCC outlays of about $49 million per year. Although the payments are made to cotton users, at least a portion of the payment is likely returned to producers in the form of higher prices associated with the increased demand from domestic users.
- Cotton Storage Cost Reimbursements. Under Section 1204(g) of the 2014 farm bill, USDA cotton storage subsidies are available to U.S. cotton producers whose cotton production is being held under the MAL program.
- STAX.[17] STAX was offered as a shallow-loss, area-wide revenue insurance product that could indemnify losses in county revenue of greater than 10% of expected revenue depending on the coverage level selected by a producer.[18] To encourage participation, USDA pays 80% of the policy premium for STAX. In contrast to the revenue guarantees available under ARC and PLC, which have a statutorily fixed lower price bound (equal to the PLC reference

[14] USDA, Farm Service Agency (FSA), "Cotton Ginning Cost-Share (CGCS) Program Fact Sheet," June 2016, https://www.fsa.usda.gov/Assets/USDA-FSA-Public/usdafiles/FactSheets/archived-fact-sheets/CGCS_Program_Fact_Sheet.pdf.

[15] FSA, "Cotton Ginning Cost-Share Fact Sheet," March 2018, https://www.fsa.usda.gov/Assets/USDA-FSA-Public/usdafiles/FactSheets/2018/cgcs_program_fact_sheet_march_2018.pdf.

[16] CBO, April 2018 baseline projections for USDA mandatory spending.

[17] See CRS Report R43494, *Crop Insurance Provisions in the 2014 Farm Bill (P.L. 113-79)*.

[18] See CRS Report R43494, *Crop Insurance Provisions in the 2014 Farm Bill (P.L. 113-79)*.

price), the revenue guarantee under STAX is recalculated each year and has decreased significantly following several years of cotton market declines since 2014. As a result, participation in STAX has been low and has declined each year that it has been available (29.3% in 2015, 25.3% in 2016, and 23.3% in 2017).

Industry Push for Cotton Reintegration into ARC/PLC Programs

Economic conditions in the U.S. farm economy steadily declined from mid-2014 into 2018, including both lower farm commodity prices and income.[19] Under these conditions, the new ARC and PLC programs paid out nearly $20 billion to eligible program crops during the first three years of the 2014 farm bill ($5.3 billion in 2014, $7.8 billion in 2015, and $6.9 billion in 2016).[20] However, cotton producers could only indirectly benefit via coupled ARC and PLC payments made to program crops planted on generic base acres. Meanwhile, net indemnity payments (indemnities minus producer-paid premium) under the new cotton revenue insurance program, STAX, were relatively small at $20.2 million in 2015 and $37.3 million in 2016, and they turned negative (i.e., premiums paid exceeded indemnities received) at -$26.1 million in 2017.[21] The National Cotton Council (NCC), in 2014, claimed that lower per-acre cotton revenues coincided with elevated costs of production and that additional economic relief was needed.[22] The U.S. cotton sector was also concerned about having resources redirected to other program crops planted on generic

---

[19] See CRS Report R45117, *U.S. Farm Income Outlook for 2018.*

[20] FSA, ARC/PLC Program website, data accessed on February 21, 2018, https://www.fsa.usda.gov/programs-andservices/arcplc_program/index.

[21] A negative net indemnity occurs when the producer-paid premiums exceed indemnity payments. USDA, Risk Management Agency (RMA), Summary of Business website, data accessed on July 12, 2018, https://prodwebnlb.rma.usda.gov/apps/SummaryOfBusiness.

[22] NCC, "NCC Thanks Lawmakers for Cottonseed Designation Push," press release, December 15, 2015, http://www.cotton.org/news/releases/2015/cotseeddesig.cfm.

base. In response to its exclusion from PLC program eligibility, the U.S. cotton sector actively sought to regain cotton's former status as a "covered commodity" but on the basis of cottonseed rather than cotton lint.

Cotton Co-Products: Lint and Cottonseed

The successful cultivation of cotton results in two products: fiber (or lint) and cottonseed, which are separated from the un-ginned cotton boll by ginning the harvested cotton. In 2017, the United States produced 9.8 billion pounds of lint[23] from upland cotton valued at $6.7 billion and 13.5 billion pounds of cottonseed[24] valued at $1 billion.[25] About 78% of U.S. upland cotton lint production is exported, while the balance is used domestically. According to the NCC, more than half of U.S. upland cotton lint production (57%) goes into apparel, 36% into home furnishings, and 7% into industrial products.[26] In contrast, cottonseed goes almost entirely to domestic users—about two-thirds is fed whole to livestock (primarily dairy cows), while the remaining seed is crushed, producing a high-grade salad oil and a high protein meal for livestock, dairy, and poultry feed.

In September 2015, the NCC proposed that the Secretary of Agriculture treat cotton lint and cottonseed separately for purposes of the ARC and PLC programs. Instead of focusing on lint, as done by past farm policy, the Secretary of Agriculture could use his apparent authority granted under the 2014 farm bill to designate cottonseed as an "other oilseed."[27] Several members of the House Agriculture Committee agreed with the NCC proposal and, in December 2015, urged then-Secretary of Agriculture Tom Vilsack to use his authority under the farm bill to

[23] Alternately, 20.570 million bales of upland cotton lint at 480 pounds per bale.

[24] Alternately, 6.725 million short tons of cottonseed at 2,000 pounds per short ton.

[25] National Agricultural Statistics Service, USDA, https://quickstats.nass.usda.gov/.

[26] NCC, "Cotton: From Field to Fabric," https://www.cotton.org/pubs/cottoncounts/fieldtofabric/upload/Cotton-FromField-to-Fabric-129k-PDF.pdf.

[27] NCC, "Developing a Cottonseed Support Program within the 2014 Farm Bill," September 21, 2015, https://www.[illegible]at.org/files/PDFs/Cottonseed-Proposal-fact-sheet%20-10-15.pdf.

designate cottonseed as a covered oilseed, thus allowing cottonseed to be a "covered commodity" eligible for the ARC and PLC programs immediately and without further action by Congress.[28]

In response, Vilsack contended that he did not have such authority because Congress had explicitly removed cotton from the list of eligible "covered commodities" and given it two special programs—STAX and CTAP, although CTAP expired in 2015. Furthermore, Vilsack argued that Congress could amend the farm bill to explicitly allow upland cotton producers to participate in ARC or PLC, or, alternatively, it could remove a provision that has appeared since FY2012 in annual appropriations acts (e.g., §715 of P.L. 115-31, the FY2017 agriculture appropriation) that effectively prohibits USDA from providing emergency assistance through the CCC.

Seed Cotton as Program Crop: Legislative Evolution

Cotton's proponents then determined that the path forward would be through an act of Congress.[29] Such an act of Congress would have budgetary implications. The general view of the agricultural community was that, if a new cotton program were to be incorporated into the next farm bill, it would likely have a budgetary cost, thus requiring possibly difficult budgetary offsets.

---

[28] See CRS Insight IN10451, *Cottonseed as a Potential Farm Program Crop: What Are the Issues?*, February 25, 2016 (available to congressional clients from the author). Furthermore, a specific provision, "Cottonseed as an Oilseed," was added in report language (CRPT 114-531) by the House appropriations subcommittee on agriculture to accompany its FY2017 appropriations bill (H.R. 5054) in April 2016.

[29] J. Coppess et al., "The Cottonseed Conundrum," *farmdoc daily*, vol. 7, no. 77 (April 27, 2017), http://farmdocdaily.illinois.edu/2017/04/the-cottonseed-conundrum.html.

PAYGO Budgetary Restrictions on Changes to the Farm Bill

Title I farm programs, in general, receive mandatory funding. Under mandatory funding, the authorizing legislation (i.e., the farm bill) defines both who is eligible to participate and how a program payment is triggered. The farm bill does not impose any spending limit on mandatory funding. Instead, program outlays are determined by producer participation, weather, and market forces. In contrast, for discretionary spending the farm bill sets program parameters and specifies funding levels. Discretionary funding levels are determined by the appropriating committees through the annual federal budget process. Thus, budget enforcement of discretionary spending is through future appropriations and budget resolutions.

Outlays for mandatory programs under a new farm bill are evaluated by the Congressional Budget Office (CBO) using multiyear "baseline" projections and are enforced through "PAYGO" budget rules.[30] CBO baseline projections involve five- and 10-year budget estimates calculated under the assumption that current law is extended over the projection period. CBO's five- and 10-year cost estimates establish a program's baseline funding.[31] The baseline is the benchmark against which proposed changes from current law that would affect mandatory spending are measured. The baseline cost impact of a proposed program change is referred to as its budget score. Any proposed program change from current law that increases outlays from CBO's baseline (such as the designation of cottonseed as an "other oilseed") that is determined to have a positive score and under PAYGO must be offset by reductions in outlays under other current-law programs to preserve baseline neutrality.

---

[30] PAYGO is a budgetary enforcement mechanism generally requiring that any direct spending and revenue legislation enacted into law may not increase the deficit. See CRS Report R41157, *The Statutory Pay-As-You-Go Act of 2010: Summary and Legislative History*, and CRS Report 97-865, *Points of Order in the Congressional Budget Process*.

[31] See CRS In Focus IF10783, *Farm Bill Primer: Budget Issues*.

FY2017 Appropriations Bill (115th Congress)

Proponents of reintegrating cotton into the ARC and PLC programs first tried to include an amendment to the FY2017 appropriations bill that would build off the 2015 NCC cottonseed proposal.[32] During spring 2017, a proposal had emerged for including cottonseed as a covered commodity with its own policy parameters—a MAL loan rate of $8.00 per cwt. and a PLC reference price of $15.88 per cwt. House negotiators identified two potential budget offsets for the cottonseed proposal: (1) declaring producers who participate in the cottonseed ARC or PLC program ineligible for STAX and (2) reallocating generic acres back to cottonseed base and away from other program crops. During the first three years of the 2014 farm bill, over $1 billion in ARC and PLC payments had been made to crops planted on generic base (*Table 1*).

Table 1. Generic Base: Planted Acres and Payments, by ARC-PLC Program

	Total Generic Base[a]	**Generic Base with Payments (1,000 Acres)**			**Payments on Generic Base ($ Million)**		
Crop Year	(1,000 Acres)	ARC	PLC	Total	ARC	PLC	Total
2014	17,583	1,372	858	2,230	$47.7	$101.2	$148.9
2015	17,583	4,871	4,833	9,704	$187.8	$256.0	$443.8
2016	17,583	4,250	4,413	8,663	$163.3	$341.7	$505.0
Total					*$398.0*	*$695.8*	*$1,097.7*

Source: FSA, ARC/PLC Program, https://www.fsa.usda.gov/programs-and-services/arcplc_program/index. Payment data for 2014 is as of September 30, 2016; 2015 data, February 2, 2017; and 2016 data, April 5, 2018.

Notes: [a] When base acres are aggregated by county by crop year, they add to 19.1 million acres in each of 2014 and 2015 as a result of double-cropping on generic base acres. FSA, "ARC/PLC Election Data," as of May 29, 2015, https://www. fsa.usda.gov/Assets/USDA-FSA-Public/usdafiles/arc-plc/excel/arc_plc_election_ data.xlsx.

[32] NCC, "Developing a Cottonseed Support Program Within the 2014 Farm Bill," September 21, 2015, https://www.ibat.org/files/PDFs/Cottonseed-Proposal-fact-sheet%20-10-15.pdf.

Thus, substantial budget offset appeared to be available. To further reduce its budgetary cost, the cottonseed proposal would not be authorized to begin or receive any funding until FY2018. Thus it would have no budget cost in the FY2017 appropriations bill under consideration.

However, the cottonseed proposal became linked to proposed changes to the dairy Margin Protection Program (MPP). Senate allies of the dairy industry sought additional baseline funding to make adjustments to the MPP in response to the dairy industry's concerns but were unable to reach a budget compromise.[33] Ultimately, the FY2017 appropriations bill that became law (Consolidated Appropriations Act of 2017, P.L. 115-31; May 5, 2017) included neither the cottonseed nor MPP proposal. Instead, the explanatory statement[34] accompanying P.L. 115-31 directed USDA to prepare a report within 60 days on the needs of cotton growers and to consider providing immediate assistance to dairy producers through the CCC.

Senate FY2018 Appropriations Bill (S. 1603, 115th Congress)

The cottonseed proposal was revised with a lower PLC reference price of $15.00 per cwt. but the same MAL loan rate of $8.00 per cwt. and was included two months later in the Senate committee-reported Agriculture appropriations bill for FY2018 (S. 1603, §728, 115th Congress, July 20, 2017).[35] The new proposal was more specific in terms of how cottonseed producers would be allowed to designate generic base acres as cottonseed base acres and about the formula to be used to determine a producer's cottonseed program yields—both of which are needed to calculate PLC

---

[33] The Senate dairy allies had offered a four-part dairy proposal that was estimated to provide $800 million in assistance over 10 years to milk producers. *Agripulse*, "Cotton-Dairy Failure Leaves a Big Budget Hole, Farm Bill Bitterness," vol. 13, no. 18 (May 3, 2017).

[34] "Division A, Agriculture, Rural Development, Food and Drug Administration, and Related Agencies Appropriations Act, 2017, Congressional Directives" for P.L. 115-31, *Congressional Record*, May 3, 2017.

[35] For a discussion of the proposed cottonseed policy parameters in the FY2018 appropriations bill, see the shaded box "Cottonseed as a Hypothetical Program Crop" and accompanying data tables in CRS Report R44914, *Farm Safety-Net Payments Under the 2014 Farm Bill: Comparison by Program Crop*.

payments. In addition, the Senate bill included dairy program changes alongside the cottonseed proposal. The joint provisions would have authority for $1 billion in new baseline spending over 10 years but required no budget offsets, because they would not start until after FY2018. However, Congress was unable to reach agreement on an FY2018 appropriations bill for reasons unrelated to agricultural policy, and the cotton issue remained unresolved into fall 2017.

Emergency Supplemental Bill (H.R. 4667, 115th Congress)

In December 2017, agricultural policy proponents in the House were able to include cotton and dairy proposals into an emergency appropriations bill (H.R. 4667) that was intended to provide assistance for victims of hurricanes and wildfires that had occurred during 2017.[36] The cotton proposal was reshaped as a "seed cotton" program that uses a weighted average of both lint and cottonseed (described below). However, the House-passed supplemental appropriations bill, H.R. 4667, was not considered in the Senate.

Bipartisan Budget Act of 2018 (P.L. 115-123; BBA)

On February 9, 2018, Congress passed a two-year budget deal—the BBA (P.L. 115-123)[37]—that included the emergency supplemental bill provisions along with both the seed cotton and dairy provisions that were in the House-passed supplemental appropriations bill (H.R. 4667). Both of these farm program changes have long-term policy implications because they change the 2014 farm bill statutes. In particular, Section 60101(a) of the BBA amends the 2014 farm bill to add seed cotton as a "covered commodity," making cotton eligible for the PLC and ARC programs.

---

[36] CRS Insight IN10843, *Supplemental Appropriations Proposed for Agriculture.*

[37] See CRS In Focus IF10829, *Agriculture Funding in the Bipartisan Budget Act of 2018.*

SEED COTTON: PROGRAM DESIGN

The BBA provision, Section 60101(a), includes seed cotton as a covered commodity, thus making it eligible for participation in PLC and ARC beginning with crop year 2018. Participating producers must make three decisions during the signup period (July 30, 2018, to December 7, 2018):[38] how to allocate their generic base acres, whether to update cotton program yields, and whether to participate in ARC or PLC. Signup is for the crop year 2018.

Allocation of Generic Base Acres

Before producers make any decision regarding ARC or PLC, they must decide how to reallocate their generic base to either seed cotton or to other program crops by the following formula:

No Recent History of Planting Covered Commodities

If the owner of a farm has not planted any covered commodities (including seed cotton) during the 2009 through 2016 crop years, then the generic base acres shall be allocated to "unassigned crop base" and no longer be eligible for ARC or PLC payments.

Recent History of Planting Covered Commodities

The owner of a farm with a history of planting covered commodities during the 2009 through 2016 crop years shall allocate generic base acres between seed cotton and other program crops as follows:

1. To seed cotton base acres in a quantity equal to the greater of 80% of their generic base or the average upland cotton plantings (or prevented from being planted) during the 2009 to 2012 crop years (not to exceed the total generic base acres on the farm) or

[38] USDA, "USDA Opens Election Period for Seed Cotton Program," press release, July 30, 2018.

2. To base acres for covered commodities (including seed cotton) in proportion to each crop's share of planted (or prevented from being planted) acreage during 2009 to 2012.[39]

Any residual or unassigned generic base acres (defined as any positive difference between generic base and the seed cotton base acres allocated under the first choice) are no longer eligible for program payments for any covered crop.

Payment Yield Choice

The seed cotton program yield is set at 2.4 times the historical upland cotton program yield (established by the 2008 farm bill [P.L. 110-246]). Producers may retain their farms' current upland cotton payment yield or opt for a one-time opportunity to update their yield based on 90% of the average yield on planted acres for the 2008 through 2012 crop years.

Program Choice

For the 2018 crop year, producers have a one-time opportunity to elect ARC or PLC for seed cotton. If no election is made, PLC becomes the automatic choice. The ARC/PLC enrollment period for the 2018 crop year began November 1, 2017, and will end on December 7, 2018.[40] It is widely assumed that cotton producers with base acres will enroll in PLC.[41]

Payment Parameters

Seed cotton uses the same payment determinations for ARC and PLC as the other program crops but subject to its own statutorily fixed reference

[39] In accordance with the reallocation formula for base acres under the 2014 farm bill and specified in code at (B),(D), (E), and (F) of 7 U.S.C. 9012(a)(3).

[40] USDA, "USDA Opens Election Period for Seed Cotton Program."

[41] J. Newton, "(Seed) Cotton Is Back in the Farm Bill," American Farm Bureau Federation, Market Intel, July 31, 2018.

price and market loan rate. The seed cotton PLC reference price is set at 36.7 cents per pound. Seed cotton payments are to be made on 85% of the seed cotton base. Because seed cotton has never been a covered commodity, no price series exists and, thus, must be defined.[42] The calculated seed cotton price (referred to as the "effective" price) is defined as the weighted annual price for upland cotton lint and cottonseed weighted by their share of combined production (lint plus cottonseed) measured in pounds (*Figure 1*). It is an artificial price that is computed only for purposes of operating the PLC and ARC programs.[43] If the effective price is less than the reference price of 36.7 cents per pound, payments are triggered under the PLC program. A producer's payment equals the product of the payment rate (i.e., the difference between the PLC reference price and the effective price) times the seed cotton program yield times 85% of the seed cotton base.

Upland cotton production remains eligible for the MAL program under the original 2014 farm bill with a loan rate that floats between 45 and 52 cents per pound of harvested cotton lint based on the simple average of the adjusted prevailing world price for the immediately preceding two market years. The new seed cotton provisions do not establish a nonrecourse MAL for seed cotton. However, for purposes of calculating the maximum potential seed cotton payment rate, seed cotton is deemed to have a MAL rate of 25 cents per pound. Finally, any farm that enrolls seed cotton base acres in either ARC or PLC is ineligible for STAX.

CBO Cost Projections

In February 2018, as part of the BBA score, CBO estimated the cost of seed cotton payments under the ARC and PLC programs at $1.270 billion

---

[42] J. Coppess et al., "Reviewing the Latest Cotton Proposal," *farmdoc daily*, vol. 8, no. 5 (January 11, 2018).

[43] In general, economists and policy analysts following this issue contend that a significant majority of cotton base owners are likely to select the PLC program (perhaps similar to the 97% or higher participation levels of canola, rice, and peanuts) based on the current outlook for potential payments under PLC. As a result, no published long-run projections of seed cotton payments under ARC have been undertaken and made public.

over five years (2018-2022) and $2.961 billion over 10 years (2018-2027).[44] However, according to CBO, the net costs of the seed cotton provision in the BBA would be reduced to $61 million over 10 years by associated budget offsets—including the reallocation of generic base and the removal of residual generic base from ARC and PLC program eligibility (-$2.188 billion over 10 years) and by repealing the eligibility for the STAX program for cotton producers that enroll their seed cotton base in either ARC or PLC (-$711 million over 10 years).

POTENTIAL ISSUES

Three issues that could figure into future policy debate as a result of this program change include the potential for large budget costs if commodity markets weaken further, the realignment of program payment acres, and considerations surrounding the potential for re-opening of the WTO U.S.-Brazil cotton dispute.

Potential Budget Cost of Seed Cotton as a Program Crop

Projections for PLC payments on seed cotton base are sensitive to changes in market prices. In April 2018 CBO released new baseline projections of USDA's mandatory farm program outlays including seed cotton as a covered commodity. Under slightly different market price expectations, CBO estimated the cost of seed cotton payments under the ARC and PLC programs to be $2.015 billion over five years (2018-2022), up 59% from the February 2018 score. *Figure 1* compares the historical and projected weighted seed cotton price since 1980 and compares it with the PLC reference price and loan rate established for seed cotton in the BBA.

---

[44] CBO score for BBA, Section 60101(a), "Treatment of Seed Cotton," February 8, 2018.

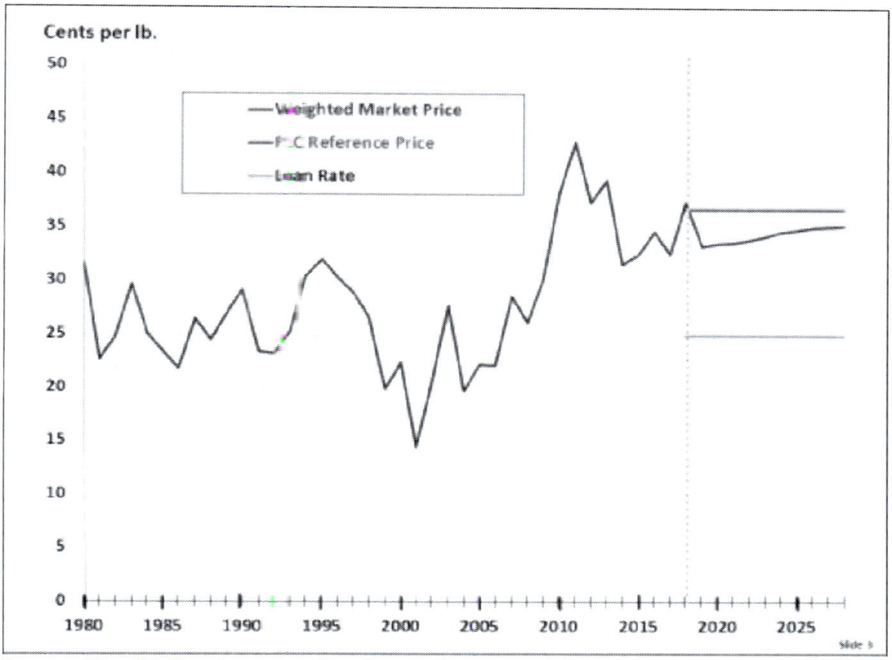

Source: The PLC reference price for seed cotton of 36.7 cents per pound, the seed cotton loan rate is 25 cents per pound, and the formula for calculating the seed cotton weighted market price ("effective price") are from the Bipartisan Budget Act of 2018 (P.L. 115-123, §60101(a)(2)). Projections for prices and production are from the CBO Baseline Projections, "USDA's Mandatory Farm Programs," April 2018.

Notes: The PLC payment rate equals the PLC reference price minus the higher of the effective price or the loan rate. The annual effective price for seed cotton is calculated as the average of the annual farm prices for upland cotton and cottonseed, each weighted by their share of total production measured in pounds. The projected annual effective price for seed cotton is calculated using price projections for upland cotton and cottonseed, weighted by share of combined production in pounds as projected by CBO.

Figure 1. PLC Reference Price Compared with Seed Cotton Weighted-Average Price (Actual and Projected).

Base Acres Reduction

The reallocation of generic base to seed cotton or other covered commodities, combined with the elimination of payments to residual base (unallocated generic base), suggests that a significant portion of historical upland cotton base acres could become ineligible for revenue support payments. The likelihood of PLC payments to seed cotton base, as suggested in the previous section, could provide an incentive for reallocating generic base to seed cotton rather than to other program crops. Upland cotton producers planted an average of 10.8 million acres during the 2009-2015 period compared with 17.6 million acres of generic base (*Figure 2*). To obtain the highest seed cotton base possible, most cotton producers might opt to reallocate 80% of generic base to seed cotton per the statutory reallocation formula for farms with a recent history of covered crops. If so, national seed cotton base would be approximately 14.1 million acres. This, in turn, would imply that as much as 20% of generic base (or 3.56 million acres) would be residual generic base and thus be disqualified from receiving farm revenue support payments. This contributes to the size of the budgetary offset in the CBO score for the BBA.

WTO U.S.-Brazil Cotton Dispute

Under the October 2014 MOU—signed by Brazil and the United States to avoid trade retaliation by Brazil—the United States agreed to limit domestic cotton support payments to those programs specifically authorized by the 2014 farm bill. However, two new U.S. cotton support programs have been implemented since the October 2014 MOU—the CGCS program and seed cotton as a covered commodity eligible for ARC and PLC payments.

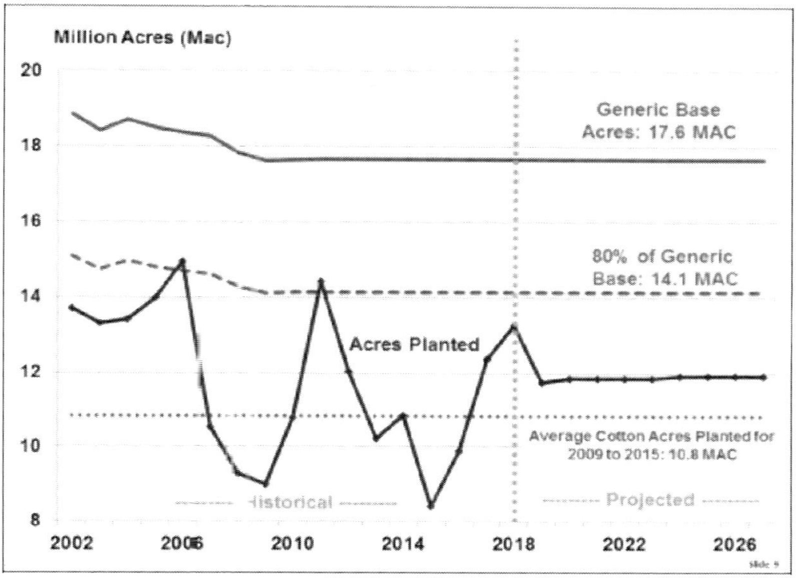

Source: Historical data for 2002-2018 are from NASS; projected data for 2019-2027 are from CBO April 2018 baseline projections.

Notes: Mac = million acres.

Figure 2. Upland Cotton: Generic Base versus Planted Acres. (Under the Provisions of the Bipartisan Budget Act of 2018, P.L. 115-123, Section 160101(a))

Potential Revival of WTO U.S.-Brazil Cotton Dispute

According to Section VI of the MOU—the Peace Clause—Brazil would refrain from retaliation prior to September 30, 2018, provided that the United States limits cotton payments to those programs specifically authorized by the 2014 farm bill. There is some ambiguity about how Brazil might respond under WTO rules, if at all, to the addition of the two new U.S. cotton programs. For example, Brazil might choose to ignore them, or it might find these new U.S. cotton support programs to be in violation of the MOU. Under the Peace Clause of the MOU,[45] Brazil must inform the United States and provide a reasonable opportunity for informal bilateral consultations before Brazil requests official dispute settlement consultations on any current domestic support program or policy.[46]

[45] WTO, "United States—Subsidies on Upland Cotton," §VI.2.

[46] WTO, "United States—Subsidies on Upland Cotton," §VI.1. (b).

Alternatively, since the two new cotton support programs were implemented after the MOU was signed, there may be a legal question as to whether these programs would qualify as "current domestic support" under the MOU. Brazil might argue that these new cotton programs fall outside the MOU and that Brazil would have no obligation to inform and informally consult with the United States before requesting official consultations as part of formal dispute settlement procedures. Or Brazil could choose to operate outside of the WTO framework and impose retaliatory tariffs (or a similar trade action) on selected U.S. agricultural products.[47]

Finally, Brazil could avoid the issue of whether the MOU has been violated by simply awaiting the MOU's expiration on September 30, 2018. There is some ambiguity about how Brazil may proceed with charges against U.S. cotton programs after the MOU's expiration—whether Brazil could use the existing dispute settlement case and extend its retaliation authority or would have to start a new WTO dispute settlement case.[48]

Potential Mitigating Factors

A potential counterargument would be that, under the seed cotton related provisions of the BBA, Congress has replaced coupled payments linked to crops planted on generic base acres with decoupled payments under the new seed cotton program, thus making U.S. farm programs more WTO-compliant, not less. It is also unclear how to interpret the CGCS repeat payments, which, although linked to the previous year's plantings, were separated by a year (2017) of non-payments. Thus, CGCS payments are not decoupled from producer behavior (potentially leading to distortive incentives), but they do not occur on a continuous basis such that cotton producers may not necessarily factor CGCS payments into their production behavior.

---

[47] See CRS Insight IN10880, *China's Retaliatory Tariffs on Selected U.S. Agricultural Products*, for an example of how China responded to U.S. Section 232 tariff actions on imports of steel and aluminum products.

[48] This report does not address any of this legal ambiguity.

Several economists have also raised more general concerns about linking cotton program payments to cotton market prices, particularly when the seed cotton reference price—which represents a composite of upland cotton and cottonseed market prices—is not intuitive and appears high relative to historical and projected market conditions.[49] Still other economists contend that large government payments tend to mute market signals, and any incentive to produce more cotton than the market can support at current prices has potentially negative impacts on developing countries.[50] However, these policy concerns may not, of themselves, be sufficient to justify a claim that the MOU has been violated or that these new programs are out of compliance with WTO rules.

Furthermore, it is not clear to what extent Brazil would have to show evidence (to either a WTO panel or the United States) that the United States is in violation of the MOU. If Brazil were compelled to show harm before a WTO panel would accept Brazil's claim that the United States had violated the MOU, this would likely delay any evidence-based determination until late 2020. This is because the seed cotton program will first be implemented for the 2018 crop. If any PLC seed cotton payments are triggered, they would not be paid out until after October 1, 2019, and would not be reported by USDA until December 2019 or January 2020 at the earliest.

---

[49] For example, see J. Coppess et al., "Reviewing the Latest Cotton Proposal," *farmdoc daily*, vol. 8, no. 5 (January 11, 2018); and J. Coppess et al., "The Cottonseed Conundrum," *farmdoc daily*, vol. 7, no. 77 (April 27, 2017).

[50] J. W. Glauber, "Unraveling Reforms? Cotton in the 2018 Farm Bill," American Enterprise Institute, January 2018.

In: Agricultural Issues
Editor: Michael Butlin

ISBN: 978-1-53615-474-0

Chapter 6

U.S. FARM INCOME OUTLOOK FOR 2018*

Randy Schnepf

SUMMARY

According to USDA's Economic Research Service (ERS), national net farm income—a key indicator of U.S. farm well-being—is forecast at $65.7 billion in 2018, down 13% from last year. This forecast does not include any payments under the trade aid package announced by USDA in July 2018. The forecast decline in 2018 net farm income is the result of lower cash returns—from both production activities (-0.1%) and government program payments (-17.4%)—and higher production expenses (up 3.3%). Net *farm* income is calculated on an accrual basis. Net *cash* income (calculated on a cash-flow basis) is also projected lower in 2018 (-12%) to $91.5 billion.

The 2018 net farm income forecast is substantially below the 10-year average of $87 billion (in nominal dollars). It would be the lowest since 2016 in nominal and inflation-adjusted 2017 dollars. This is primarily the result of the outlook for continued weak prices for most major crops. U.S. farm income experienced a golden period during 2011 through 2014 due to strong commodity prices and robust agricultural exports. Most crops

* This is an edited, reformatted and augmented version of Congressional Research Service Publication No. R45117, Updated September 21, 2018.

and livestock product prices remain significantly below the average for 2011-2014, when prices for many major commodities attained record or near-record highs. Net farm income is projected down 47% since its record high of $123.7 billion in 2013. Net cash income is projected down 32% from its 2013 high of $135.6 billion. Farm-sector production expenses have risen 1.6% over that same period, thus contributing to lower aggregate income totals.

Government payments are projected down 17% from 2017 at $9.3 billion—due largely to projected lower payments in 2018 under the Price Loss Coverage (PLC) and Agricultural Risk Coverage (ARC) revenue support programs for major field crops (projected at $3.5 billion, down 50% from $7.0 billion in 2017).

Since 2008, U.S. agricultural exports have accounted for a 20% share of U.S. farm and manufactured or processed agricultural sales. In 2018 agricultural exports are forecast to be flat, at $144.5 billion, due largely to abundant supplies in international markets and strong competition from major foreign competitors. The ongoing U.S.-China trade dispute is expected to shift trade patterns, but the aggregate value of U.S. exports is expected to hold steady.

In spite of the lower farm income outlook, farm wealth is projected up 1.2% from 2017 to $3.0 trillion. Farm asset values reflect farm investors' and lenders' expectations about long-term profitability of farm sector investments. Farmland values are projected up 1.9% in 2018 following a 2.3% rise in 2017. Because they comprise such a significant portion of the U.S. farm sector's asset base (81%), change in farmland values is a critical barometer of the farm sector's financial performance. At the farm household level, average farm household incomes have been well above average U.S. household incomes since the late 1990s. In 2016 (the last year with comparable data), the average farm household income (including off-farm income sources) of $117,918 was about 42% higher than the average U.S. household income of $83,143.

The outlook for lower net farm income and relatively weak prices for most major program crops signals the likelihood of continued relatively lean times ahead. With record soybean and near-record corn crops predicted by USDA, the U.S. agricultural sector's well-being remains dependent on continued growth in domestic and foreign demand to sustain prices at current modest levels. In addition to commodity prices, the financial picture for the agricultural sector as a whole heading into 2019 will hinge on both domestic and international macroeconomic factors, including interest rates, economic growth, and consumer demand. This chapter incorporates USDA's August 30, 2018, farm income projections and its August 29, 2017, U.S. agricultural trade outlook update.

INTRODUCTION

The U.S. farm sector is vast and varied. It encompasses production activities related to traditional field crops (such as corn, soybeans, wheat, and cotton) and livestock and poultry products (including meat, dairy, and eggs), as well as fruits, tree nuts, and vegetables. In addition, U.S. agricultural output includes greenhouse and nursery products, forest products, custom work, machine hire, and other farm-related activities. The intensity and economic importance of each of these activities, as well as their underlying market structure and production processes, vary regionally based on the agro-climatic setting, market conditions, and other factors. As a result, farm income and rural economic conditions may vary substantially across the United States.

Annual U.S. net farm income is the single most watched indicator of farm sector well-being, as it captures and reflects the entirety of economic activity across the range of production processes, input expenses, and marketing conditions that have prevailed during a specific time period. When national net farm income is reported together with a measure of the national farm debt-to-asset ratio, the two summary statistics provide a quick and widely referenced indicator of the economic well-being of the national farm economy.

Measuring Farm Profitability

Two different indicators measure farm profitability: net cash income and net farm income.

Net cash income compares cash receipts to cash expenses. As such, it is a cash flow measure representing the funds that are available to farm operators to meet family living expenses and make debt payments. For example, crops that are produced and harvested but kept in on-farm storage are not counted in net cash income. Farm output must be sold before it is counted as part of the household's cash flow.

Net farm income is a more comprehensive measure of farm profitability. It measures value of production indicating the farm operator's share of the net value added to the national economy within a calendar year independent of whether it is received in cash or noncash form. As a result, net farm income includes the value of home consumption, changes in inventories, capital replacement, and implicit rent and expenses related to the farm operator's dwelling that are not reflected in cash transactions. Thus, once a crop is grown and harvested, it is included in the farm's net income calculation, even if it remains in on-farm storage.

Key Concepts

- Net cash income is generally less variable than net farm income. Farmers can manage the timing of crop and
- livestock sales and purchase of inputs to stabilize the variability in their net cash income. For example, farmers can hold crops from large harvests to sell in the forthcoming year, when output may be lower and prices higher.
- Off-farm income and crop insurance subsidies, both of which have increased in importance in recent years,
- are not included in the calculation of aggregate farm income. Crop insurance indemnity payments are included.
- Off-farm income is included in the discussion of farm income at the household level at the end of this chapter.

National versus State-Level Farm Household Data

This chapter focuses singularly on aggregate national net farm income projections for 2018 as reported by the U.S. Department of Agriculture's (USDA) Economic Research Service (ERS).[1] Aggregate data often hide or understate the tremendous diversity and regional variation that occurs across America's agricultural landscape. For insights into the differences in American agriculture, visit the ERS websites on "Farm Structure and Organization" and "Farm Household Well-being.[2]"

---

[1] For both national and state-level farm income, see ERS, "U.S. and State Farm Income and Wealth Statistics," http://www.ers.usda.gov/data-products/farm-income-and-wealth-statistics.aspx.

[2] ERS, "Farm Structure and Organization," http://www.ers.usda.gov/topics/farm-economy/farm-structure-andorganization.aspx; and "Farm Household Well-Being," http://www.ers.usda.gov/topics/farm-economy/farm-householdwell-being.aspx

USDA's 2018 Farm Income Forecast

According to ERS, U.S. net farm income is forecast at $65.7 billion in 2018, down 13% from last year (Table A-1).[3] This forecast does not include any payments under the trade aid package announced by USDA in July 2018.[4] Under the initial phase of the trade aid package, USDA expects to provide $4.7 billion in direct payments to qualifying agricultural producers.

The forecast decline in 2018 net farm income represents continued agriculture-sector economic weakness since 2013's record high of $123.8 billion. Abundant domestic and international supplies of grains and oilseeds suggest a fourth straight year of relatively weak commodity prices in 2018.

Selected Highlights

- The net farm income forecast for 2018 is substantially below the 10-year average of $87 billion (Figure 1). In both nominal and inflation-adjusted dollars, the 2018 forecast is the lowest since 2016 (Figure 2).
- Net cash income is also projected lower in 2018 at $91.5 billion (-12%). Since the record highs of 2012 and 2013, net cash income and net farm income have fallen by 32% and 47%, respectively (Figure 1), thus continuing a general downward trend in farm income since 2013—primarily the result of a significant decline in most farm commodity prices since the 2013-2014 period (Figure A-1 to Figure A-4).
- Cash receipts for most major field crops (feed grains, hay, and wheat), oilseeds (Figure 12), and animal products (beef, pork,

[3] The material presented in the report is drawn primarily from the 2018 Farm Sector Income Forecast of ERS at https://www.ers.usda.gov/topics/farm-economy/farm-sector-income-finances/farm-sector-income-forecast/.

[4] CRS Report R45310, *Farm Policy: USDA's Trade Aid Package*, by Randy Schnepf et al.

broilers, eggs, and milk; Figure 14) have declined since 2013-2014 as U.S. and global grain and oilseed stocks and animal herds have rebuilt (Table A-4).

- Government payments in 2018 are projected down (-17.4%) from 2017 at $9.5 billion (Figure 15). Under the Agricultural Risk Coverage (ARC) program, several years of low farm prices have reduced the historical revenue guarantee and resulting payments, while Price Loss Coverage (PLC) program payments are projected lower due to slightly stronger farm prices compared with 2017.
- Total production expenses for 2018 (Figure 16), at $365.9 billion, are projected up from 2017 (+3.3%), driven largely by feed, fuel, labor, and interest costs.
- Global demand for U.S. agricultural exports (Figure 20) is expected to surpass 2017's relatively strong value of $144 billion in 2018 by $0.5 billion—still below the record of $152.3 billion set in 2014.[5]
- Farm asset values and debt levels are projected to be up slightly in 2018—asset values at $3.0 trillion (+1.2%) and farm debt at $407 billion (+3.5%)—pushing the projected debt-to-asset ratio up to 13.4%, the highest level since 2003 (*Figure 26*).

U.S. Agriculture Outlook: 2018 Overview

In January 2018, USDA reported 2017 fall winter wheat plantings (which would grow out as part of the 2018 wheat crop) at 32.7 million acres—the lowest level since 1909.[6] The low winter wheat planted acres reflected poor profitability when compared with soybeans. U.S. soybean plantings have been steadily expanding out of the traditional Corn Belt—the fertile agricultural zone stretching from Ohio westward through Indiana, Illinois, Iowa, southern Minnesota, and northern Missouri to the

[5] ERS, Outlook for U.S. Agricultural Trade, AES-105, August 29, 2018.

[6] USDA, National Agricultural Statistics Service (NASS), *Winter Wheat and Canola Seedings*, January 12, 2018.

eastern half of Nebraska and Kansas—and into the Central and Northern Plains.

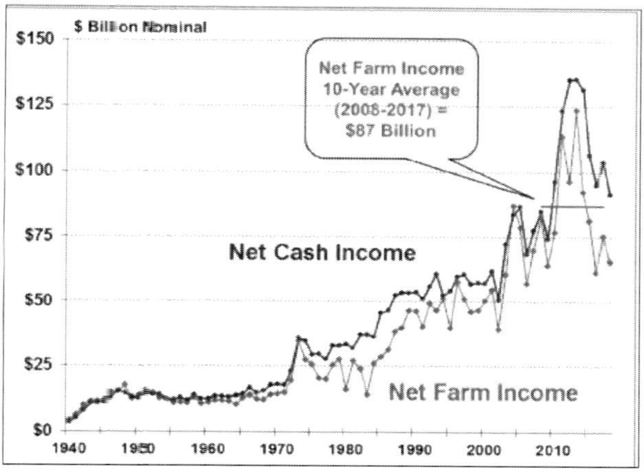

Source: ERS, "2018 Farm Income Forecast," August 30, 2018. All values are nominal—that is, not adjusted for inflation. Values for 2018 are forecasts.

Figure 1. Annual U.S. Farm Sector Nominal Income, 1940-2018.

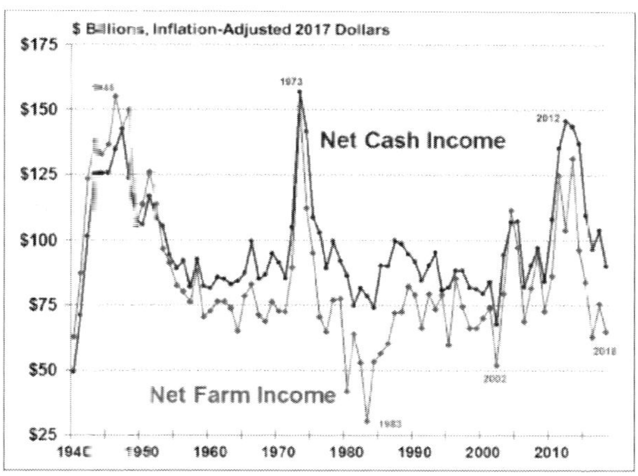

Source: ERS, "2018 Farm Income Forecast," August 30, 2018. All values are adjusted for inflation using the chain-type gross domestic product deflator, where 2017 = 100, Office of Management and Budget, Historical Tables, Table 10.1, https://www.whitehouse.gov/omb/budget/Historicals; 2018 is forecast.

Figure 2. Annual U.S. Farm Sector Inflation-Adjusted Income, 1940-2018.

Heading into the 2018 spring planting period, USDA weather reports in early February indicated increasingly dry conditions across the southern tier of the United States and the northern plains (Figure 3). Those dry conditions persisted through much of the growing season (Figure 4). Such dry conditions generated considerable concerns for producers with no or little access to irrigation. They are also of significant concern to the U.S. cattle sector, which depends on spring and early summer grazing across the Northern and Southern Plains and on cheap and abundant feed grains through the remainder of the year. In contrast to the Plains states, the major growing zones of the Corn Belt and Southeast received favorable weather through the growing season, contributing to USDA's forecast for large harvests of corn, soybeans, rice, and cotton in 2018.

Large Corn and Soybean Crop Prospects Dominate Markets

Corn and soybeans are the two largest U.S. commercial crops in terms of both value and quantity. For the past several years, U.S. corn and soybean crops have experienced remarkable growth in both productivity and output, thus helping to build stockpiles at the end of the marketing year. This has been particularly true for soybeans, which has seen rapid growth in yield, acres planted, and stocks (Figure 6). This pattern continued into 2018 when, for the first time in history, U.S. soybean plantings (at 89.6 million acres) exceeded corn plantings (89.1 million acres). The record soybean plantings coupled with record yield (52.8 bushels/acres) is expected to produce a record U.S. soybean harvest of 4.7 billion bushels in 2018.[7] The record harvest is expected to pressure soybean farm prices lower (-8%) to a projected $8.60 per bushel for the 2018/19 marketing year (Figure 5).[8]

For corn, USDA forecasts a record national average yield of 181.3 bushels/acre (up 2.7% from the previous year's record yield). As a result,

---

[7] World Agricultural Outlook Board (WAOB), USDA, *World Agricultural Supply and Demand Estimates (WASDE)*, September 12, 2018.

[8] The marketing year varies slightly with each agricultural commodity, but it tends to begin at harvest and end before the next year's harvest. For example, the marketing year for corn planted and harvested in 2018 begins September 1, 2018, and ends August 31, 2019.

the United States s expected to produce the second largest corn harvest on record at 14.8 billion bushels.

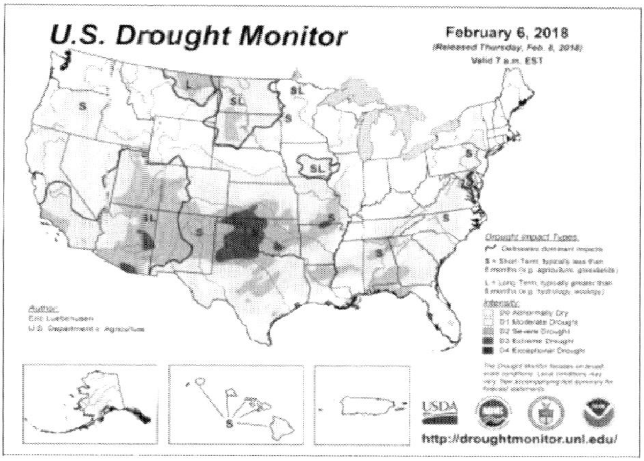

Source: USDA, Office of the Chief Economist, Current U.S. Drought Monitor, February 6, 2018. http://droughtmonitor.unl.edu/data/jpg/current/current_usdm.jpg.

Figure 3. Dry Conditions Across South, and North-Central Heading into 2018.

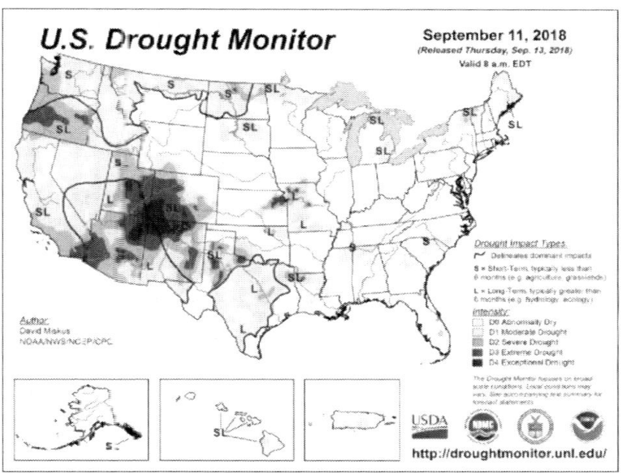

Source: USDA, Office of the Chief Economist, Current U.S. Drought Monitor, September 11, 2018, http://droughtmonitor.unl.edu/data/jpg/current/current_usdm.jpg.

Figure 4. Dry Conditions Across Southern Plains, Parts of Corn Belt.

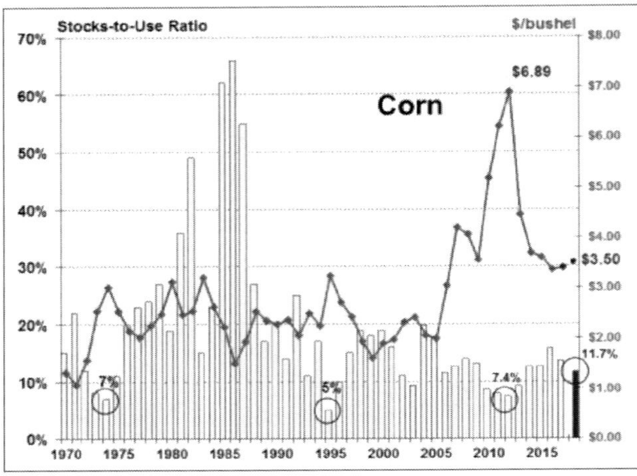

Source: World Agricultural Outlook Board (WAOB), USDA, World Agricultural Supply and Demand Estimates (WASDE), September 13, 2018. All values are nominal. Values for 2018 are forecasts.

Notes: Stocks-to-Use equals the ratio of season-ending stocks relative to the season's total usage.

Figure 5. U.S. Corn Stocks-to-Use Ratio Down, Price Up Slightly in 2018.

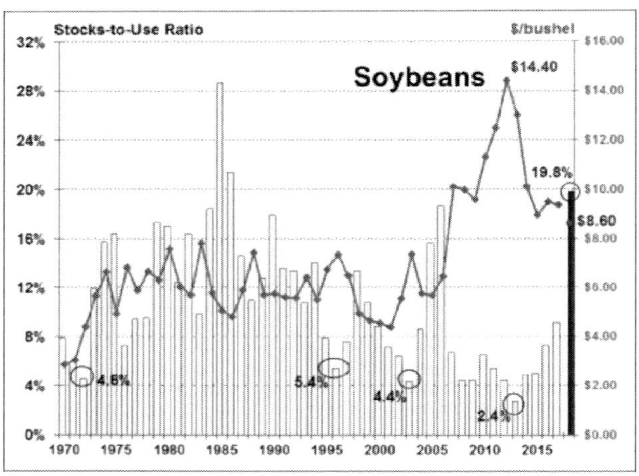

Source: WAOB, USDA, WASDE, September 13, 2018. All values are nominal. Values for 2018 are forecasts.

Notes: Stocks-to-Use equals the ratio of season-ending stocks relative to the season's total usage.

Figure 6. U.S. Soybean Stocks-to-Use Ratio Up Sharply, Price Down in 2018.

Despite the record production, USDA predicts that record large domestic usage (including for livestock feed, ethanol production, other industrial processing, and seed) plus continued strong exports (third-largest on record at 2.4 billion bushels) will result in a reduction in corn ending stocks and a slightly higher season average farm price of $3.50/bushel (up 2.9% year-over-year).

Early Livestock Outlook for 2018

USDA's January 2018 *Cattle* report reported that U.S. cattle herd expansion, which has been growing since 2015, has slowed markedly but is still projected to grow through 2018.[9] Similarly, U.S. hog and poultry flocks have been growing and are expected to continue to expand in 2018.[10]

Record profitability among cow-calf producers in 2014 coupled with then-improved forage conditions helped to set off the beef cow herd expansion (Figure 7).[11] The continued cattle expansion through 2018—despite weakening profitability—is primarily the result of a substantial lag in the biological response to the strong market price signals of late 2014. Unfortunately, drought-like conditions have persisted throughout the spring and summer months in the Southern Plains where much of U.S. cattle production occurs (Figure 4). Dry pasture and range conditions tend to limit feeding choices and raise costs for ranchers in affected areas.

During the 2007-2014 period, high feed and forage prices plus widespread drought in the Southern Plains—the largest U.S. cattle production region—had resulted in an 8% contraction of the U.S. cattle inventory (Figure 7). Reduced beef supplies led to higher producer and consumer prices, which in turn triggered the slow rebuilding phase in the cattle cycle that started in 2014 (see the price-to-feed ratio for steers and heifers, Figure 8). The resulting continued expansion of beef supplies pressured market prices lower in 2016. The lower price outlook is expected

---

9 NASS, *Cattle*, January 31, 2018.

10 WAOB, USDA, *WASDE*, Table—U.S. Quarterly Animal Product Production, September 12, 2018, p. 31.

11 J. Mintert, "Cattle Inventory Growth Slowing Down, but Beef Production Still Increasing," *farmdoc daily*, vol. 8, no. 18 February 5, 2018.

to persist through 2019, despite strong domestic and international demand across all meat categories—beef, pork, and poultry (*Table A-4*).

For 2018, increases in production of beef (+3.5%), pork (+4.3%),[12] broilers (+2.3%), and eggs (+1.7%) are projected to expand relatively robustly. Fortunately for producers, USDA projects that combined domestic and export demand will continue to grow for red meat (+4.4%) and poultry (+2.2%), thus helping to support prices and profit margins in 2018. Production is projected to grow in 2019 but at slightly lower rates: beef (+2.3%), pork (+4.5%), broilers (+1.9%), and eggs (+1.9%).

The changing conditions for the U.S. livestock sector may be tracked by the evolution of the ratios of livestock output prices to feed costs (*Figure 8*). A higher ratio suggests greater profitability for producers.[13] The cattle-, hog-, and broiler-to-feed margins all moved upwards in during 2014 but have exhibited substantial volatility during the 2015-2018 period.[14] The hog, cattle, and broiler sectors remain profitable. However, continued production growth of between 2% and 3% for red meat and poultry suggests that prices are vulnerable to weakness in demand. Milk prices and the milk-to-feed ratio fell sharply during 2018, suggesting eroding profitability. While this result varies widely across the United States, many small or marginally profitable producers face continued financial difficulties. In addition, both U.S. and global milk production are projected to continue growing in 2018. As a result, milk prices could come under further pressure in 2019; although USDA is currently projecting milk prices up slightly in 2019.[15]

With respect to the federal milk margin protection program (MPP) instituted by the 2014 farm bill (Agricultural Act of 2014, P.L. 113-79), the formula-based milk-to-feed margin used to determine government

---

[12] USDA's August 30, 2018, farm income projections do not consider the effects of Hurricane Florence on North Carolina hog and poultry production.

[13] Feed costs—at 30% to 80% of variable costs—are generally the largest cost component in livestock operations.

[14] Broilers are chickens raised for meat and contrasts with layers, which are chickens retained for egg production.

[15] WAOB, USDA, *WASDE*, Table—U.S. Quarterly Prices for Animal Products, September 12, 2018, p. 31.

payments has fallen below the $8.00 per hundredweight (cwt.) threshold needed to trigger payments in early 2018 (*Figure 9*).[16]

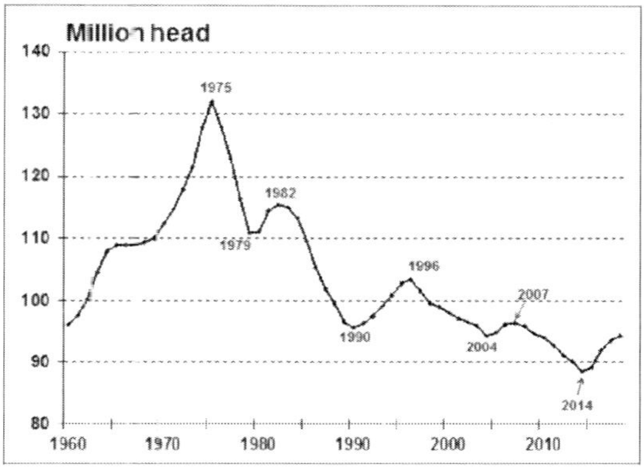

Source: USDA, National Agricultural Statistics Service (NASS), Cattle, January 31, 2018.

Notes: Inventory data are for January 1 of each year.

Figure 7. The U.S. Beef Cattle Inventory (Including Calves) Since 1960.

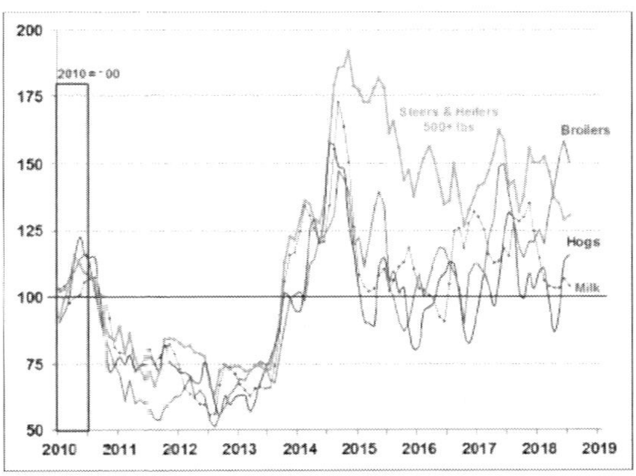

Source: NASS, Agricultural Prices, August 29, 2018.

Notes: Cattle and hog feed cost is 100% corn. Broilers feed cost is 58% corn and 42% soybeans. Dairy feed cost is a mix of corn, soybean meal, and alfalfa hay.

Figure 8. Indexed Farm-Price-to-Feed Ratios for Cattle, Broilers, Milk, and Hogs (Ratio of national average farm price per 100 lbs. of meat to per-unit feed cost. Indexed, 2010 = 100).

[16] See CRS In Focus IF10195, *U.S. Dairy Programs After the 2014 Farm Bill (P.L. 113-79).*

The MPP margin differs from the USDA-reported milk-to-feed ratio shown in Figure 8 but reflects the same market forces.

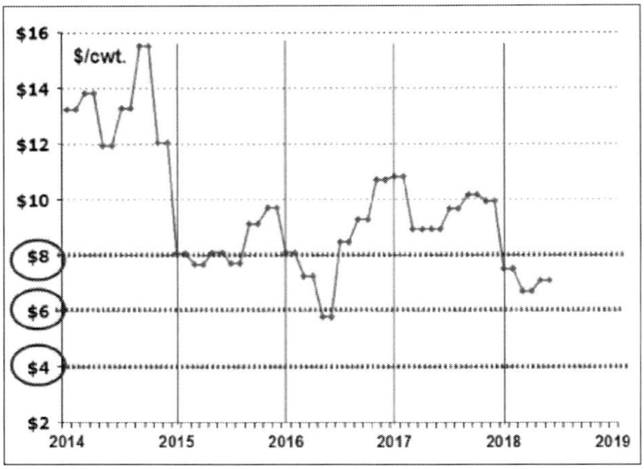

Source: NASS, Agricultural Prices, August 30, 2018; calculations by CRS. All values are nominal.

Note: The margin equals the All Milk price minus a composite feed price based on the formula used by the MPP of the 2014 farm bill (P.L. 113-79). See CRS Report R43465, Dairy Provisions in the 2014 Farm Bill (P.L. 113-79).

Figure 9. The MPP Margin Has Fallen Below $8/cwt. in Early 2018. (National average farm price of milk less average feed costs per 100 lbs.)

USDA Trade aid Package

The corn and soybean crops provide important inputs for the domestic livestock, poultry, and biofuels sectors. In addition, the United States is traditionally one of the world's leading exporters of corn, soybeans, and soybean products—vegetable oil and meal. During the recent five-year period from 2013/2014 to 2017/2018, the United States exported 49% of its soybean production and 15% of its corn crop. As a result, the outlook for these two crops is critical to both farm sector profitability and regional economic activity across large swaths of the United States as well as in international markets. However, a tariff-related trade dispute between the

United States and several major trading partners has cast considerable uncertainty over the corn and soybean markets.[17]

As a result of the trade dispute, in July 2018, USDA announced a trade aid package including up to $12 billion in financial support designed to partially offset the negative price and income effects of lost commodity sales to major markets.[18] In August 2018, Secretary Perdue announced details of the trade aid package that included initial direct payments of $4.7 billion to qualifying producers of seven agricultural commodities: soybeans, corn, cotton, sorghum, wheat, hogs, and dairy. The August trade aid package also included $1.2 billion in government purchases of excess food supplies and $200 million in trade promotion. The USDA trade aid announcement came too late to be included in USDA's August 30 farm income forecast shown in this chapter. However, their effect would be to increase net farm income for 2018 by an equivalent amount to $70.4 billion assuming that the entire $4.7 billion is distributed in 2018.

Gross Cash Income Highlights

Total farm sector gross cash income for 2018 is projected to be down (-0.1%) to $418.2 billion, driven by a decline (-0.1%) in the value of agricultural sector production (Figure 10), including a small decline in the value of crop output (-0 3%) partially offset by a slight rise in livestock output (0.1%). An increase in projected farm-related income (+5.6%) is offset by sharply lower government payments (-17.4%) (Table A-1).

Projected farm-sector revenue sources and revenue shares include crop revenues (47% of sector revenues), livestock receipts (42%), government payments (2%), and other farm-related income, including crop insurance indemnities, machine hire, and custom work (8%).

---

[17] CRS Insight IN10943, Escalating Tariffs: Timeline and Potential Impact, coordinated by Brock R. Williams; and CRS Insight IN10880, China's Retaliatory Tariffs on Selected U.S. Agricultural Products, by Jenny Hopkinson.

[18] CRS Report R45310, *Farm Policy: USDA's Trade Aid Package*, by Randy Schnepf et al.

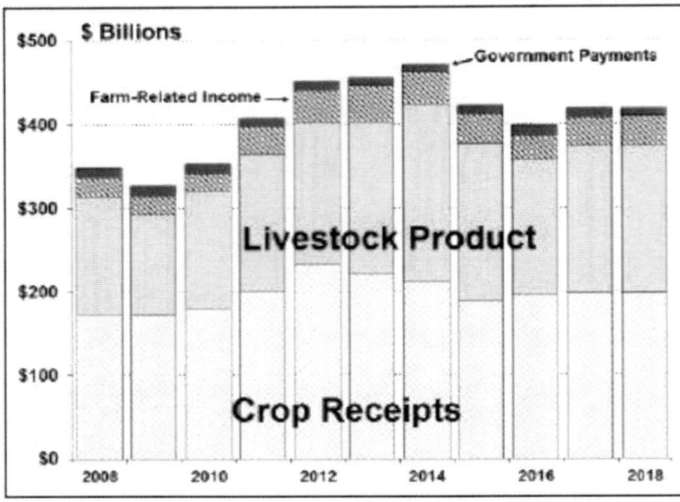

Source: ERS, "2018 Farm Income Forecast," August 30, 2018. All values are nominal. Values for 2018 are forecasts

Note: Reciepts frm crop and livestock product sales, and government payments, are described in "Gross Cash Income Highlights." Farm-related income includes income from custom work, machine hire, agro-tourism, forest product sales, insurance indemnitites, and cooperative patronage dividend fees.

Figure 10. Farm cash receipts by source, 2008-2018.

Crop Receipts

Total crop sales peaked in 2012 at $231.6 billion when a nationwide drought pushed commodity prices to record or near-record levels. In 2018, crop sales are projected at $197.8 billion, down slightly from 2017 (Figure 11). The crop sector includes 2018 projections (and percentage changes from 2017) for the following commodity groups:

- Feed crops—corn, barley, oats, sorghum, and hay: $55.2 billion (-0.6%);
- Oil crops—soybeans, peanuts, and other oilseeds: $43.2 billion (+0.4%);
- Food grains—wheat and rice: $11.5 billion (+4.4%);
- Fruits and nuts: $30.4 billion (-2.2%);
- Vegetables and melons: $19.1 billion (-3.8%);

- Cotton: $7.9 billion (+1.0%); and
- All other crops—including tobacco, sugar, greenhouse, and nursery crops: $29.2 billion (+1.8%).

Livestock Receipts

The livestock sector includes cattle, hogs, sheep, poultry and eggs, dairy, and other minor activities. Cash receipts for the livestock sector grew steadily from 2009 to 2014, when it peaked at a record $212.8 billion. However, the sector turned downward in 2015 (-11.0%) and again in 2016 (-14.1%), driven largely by projected year-over-year price declines across major livestock categories (*Table A-4* and *Figure 13*). In 2017, livestock sector cash receipts recovered with year-to-year growth of 8.1% to $176.0 billion. In 2018, strong cash receipts for the poultry and egg sector are projected to offset declines in cattle, hog, and dairy sales, keeping overall livestock cash receipts flat at $176.2 billion (+0.1%). Highlights include 2018 projections (and percentage changes from 2017) for the following livestock groups:

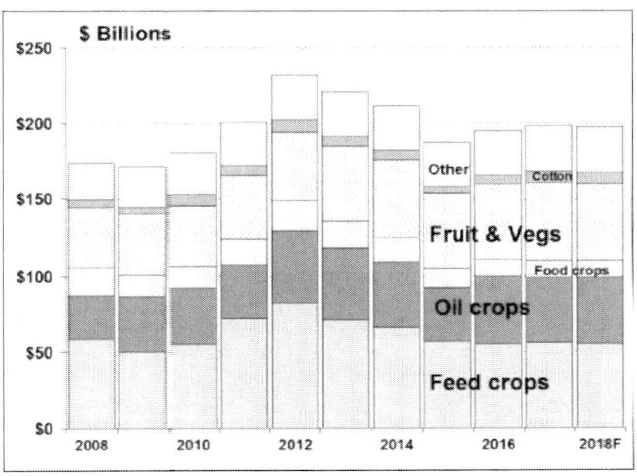

Source: ERS, "2018 Farm Income Forecast," August 30, 2018. All values are nominal. Values for 2018 are forecasts.

Figure 11. Crop Cash Receipts by Source, 2008-2018.

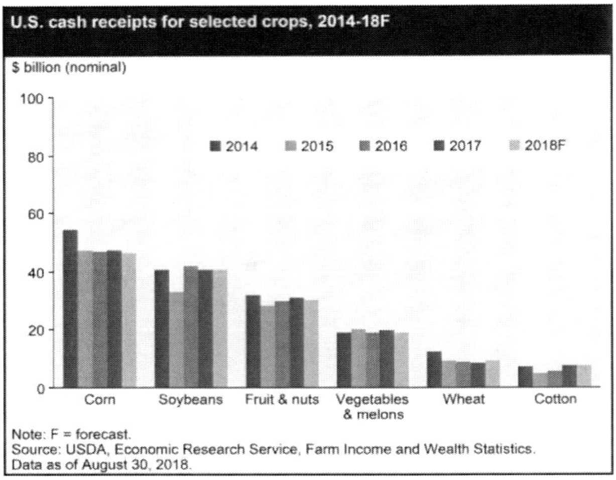

Source: ERS, "2018 Farm Income Forecast," August 30, 2018. All values are nominal. Values for 2018 are forecasts.

Figure 12. Cash Receipts for Selected Crops, 2014-2018.

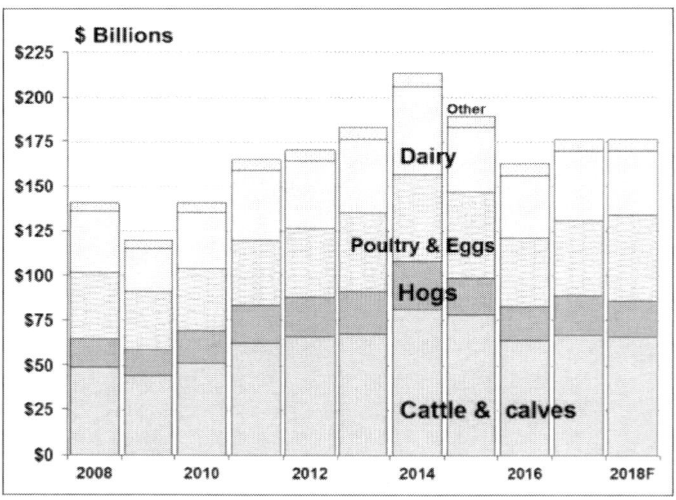

Source: ERS, "2018 Farm Income Forecast," August 30, 2018. All values are nominal. Values for 2018 are forecasts.

Figure 13. U.S. Livestock Product Cash Receipts by Source, 2008-2018.

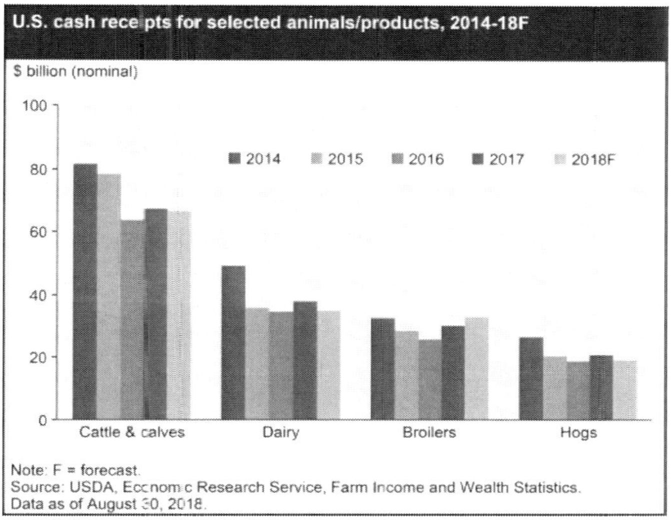

Source: ERS, "2018 Farm Income Forecast," August 30, 2018. All values are nominal. Values for 2018 are forecasts.

Figure 14. Cash Receipts for Selected Animal Products, 2014-2018.

- Cattle and calf sales: $66.6 billion (-1.1%),
- Hog sales: $19.4 billion (-7.7%),
- Poultry and egg sales: $47.8 billion (+12.1%),
- Dairy sales, valued at $35.1 billion (-7.4%), and
- Miscellaneous livestock,[19] valued at $7.2 billion (+3.1%).

Government Payments

Government payments in 2018 are projected down 18.6% from 2017 at $9.5 billion due largely to projected lower payments under the Price Loss Coverage (PLC) and Agricultural Risk Coverage (ARC) revenue support programs for major field crops—down to $3.5 billion from $7.0 billion in 2017 (See "Price Contingent" in Figure 15).[20]

---

[19] Miscellaneous livestock includes aquaculture, sheep and lambs, honey, mohair, wool, pelts, and other animal products.

[20] For details see CRS Report R43448, Farm Commodity Provisions in the 2014 Farm Bill (P.L. 113-79).

Government payments of $9.5 billion would represent a relatively small share (2%) of projected gross cash income of $418.2 billion in 2018. In contrast, government payments are expected to represent 14% of the projected net farm income of $65.7 billion. However, the importance of government payments as a percent of net farm income varies nationally by crop and livestock sector and by region.

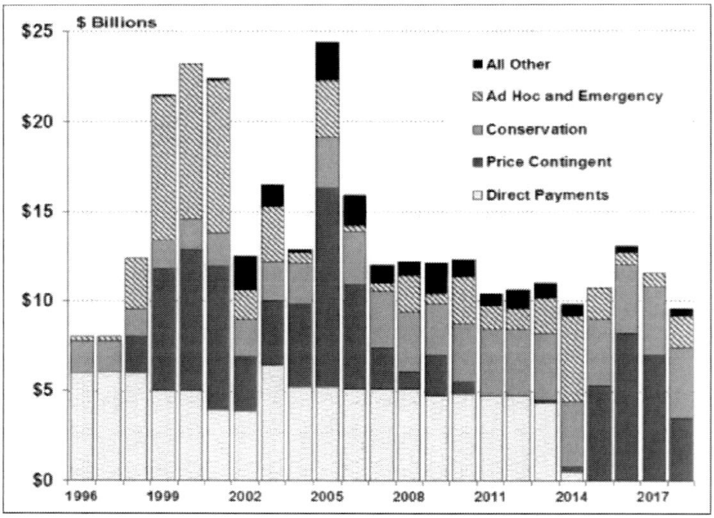

Source: ERS, "2018 Farm Income Forecast," August 30, 2018. All values are nominal. Values for 2018 are forecasts.

Notes: Data are on a fiscal year basis and may not correspond exactly with the crop or calendar year. "Direct Payments" include production flexibility contract payments enacted under the 1996 farm bill and fixed direct payments of the 2002 and 2008 farm bills. "Price-Contingent" outlays include loan deficiency payments, marketing loan gains, CCP, ACRE, PLC, and ARC payments. "Conservation" outlays include Conservation Reserve Program payments along with other conservation program outlays. "Ad Hoc and Emergency" includes emergency supplemental crop and livestock disaster payments and market loss assistance payments for relief of low commodity prices. "All Other" outlays include cotton ginning cost-share, biomass crop assistance program, peanut quota buyout, milk income loss, tobacco transition, and other miscellaneous payments.

Figure 15. U.S. Government Farm Support, Direct Outlays, 1996-2018.

No payments are forecast under the marketing loan program in 2018, compared with $11.9 million in 2017 and $206 million in 2016, as program crop prices are expected to remain above most program loan rates through 2018 (Table A-4). The MPP for dairy is expected to make $218 million in MPP payments in 2018. Conservation programs include all conservation programs operated by USDA's Farm Service Agency and the Natural Resources Conservation Service that provide direct payments to producers.

Estimated conservation payments of $3.9 billion are forecast for 2018, up slightly from $3.8 billion in 2017. USDA forecasts supplemental and ad hoc disaster assistance payments at $1.655 billion in 2018, up from $679 million in 2017.[21]

Production Expenses

Total production expenses for 2018 for the U.S. agricultural sector are projected to be up (+3.3%) from 2017 in nominal dollars at $365.9 billion (Figure 16). A comparison of the indexes of prices paid versus prices received reveals that the prices received index generally declined from 2014 through 2016, rebounded in 2017, then declined again in 2018 (Figure 17). Farm input prices (as reflected by the prices paid index) showed a similar pattern but with a much smaller decline from their 2014 peak, thus suggesting that farm sector profit margins have been squeezed since 2014.

Production expenses will affect crop and livestock farms differently. The principal expenses for livestock farms are feed costs, purchases of feeder animals and poultry, and hired labor. Feed costs, replacement animal costs, interest costs, and hired labor are projected up in 2018 (Figure 18). Projected changes to the principal crop expenses in 2018 are mixed: fuel and land rent are projected up, while fertilizers costs are projected lower.

---

[21] Fiscal year payments generally involve outlay commitments incurred during the previous crop year. For example, FY2018 disaster assistance payments are primarily related to disasters for crops that were grown and harvested in 2017. See CRS Report RS21212, *Agricultural Disaster Assistance*, for information on available farm disaster programs.

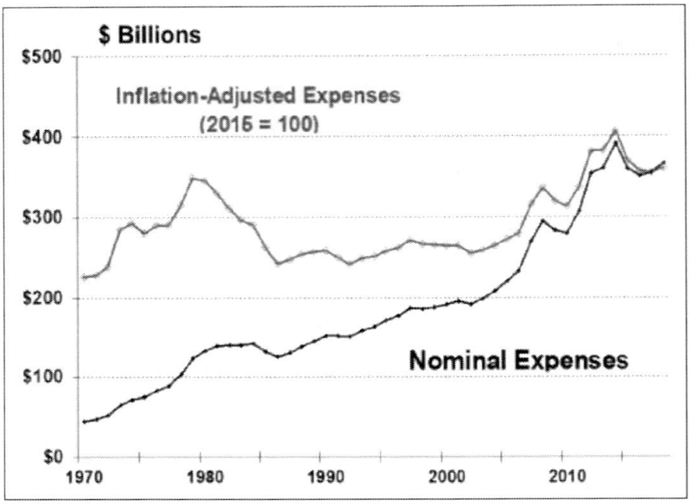

Source: ERS, "2018 Farm Income Forecast," August 30, 2018. Inflation-adjusted expenses are calculated using the chain-type GDP deflator, OMB, Historical Tables, Table 10.1. Amounts for 2018 are forecasts.

Figure 16. Total Annual Farm Production Expenses, 1970-2018.

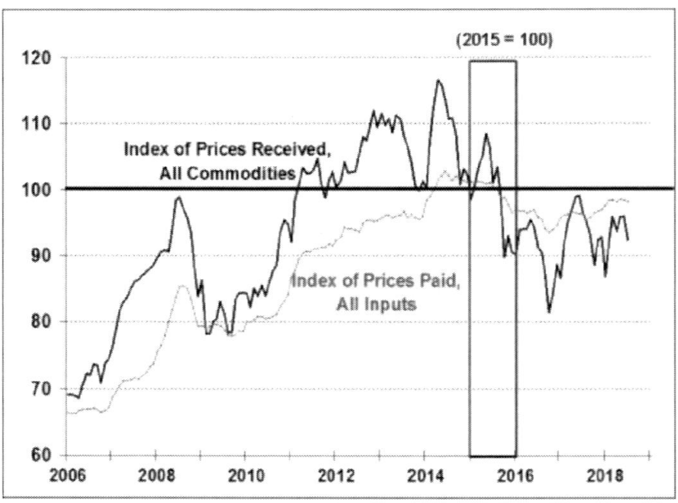

Source: NASS, Agricultural Prices, August 30, 2018.

Notes: Prices are indexed to 2015 = 100 to permit relative comparisons. Calculations by CRS.

Figure 17. Index of Monthly Prices Received versus Prices Paid, 2006-2018.

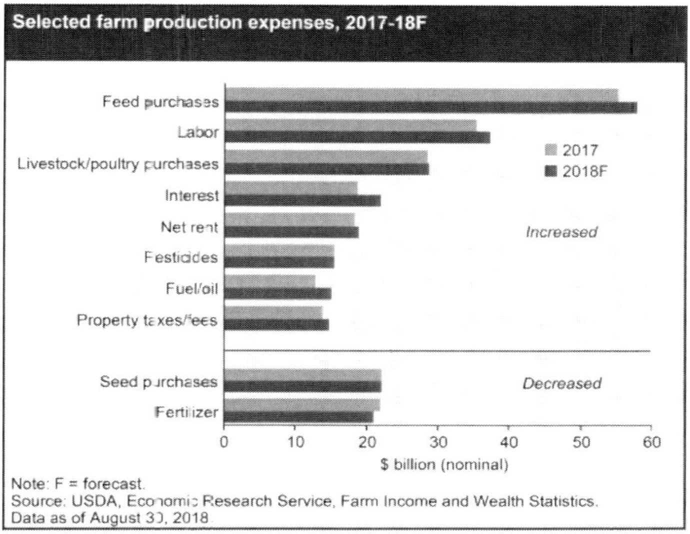

Source: ERS, "2018 Farm Income Forecast," August 30, 2018. All values are nominal. Amounts for 2018 are forecasts.

Figure 18. Farm Production Expenses for Selected Items, 2017 and 2018.

Cash Rental Rates

Renting or leasing land is a way for young or beginning farmers to enter agriculture without incurring debt associated with land purchases. It is also a means for existing farm operations to adjust production more quickly in response to changing market and production conditions while avoiding risks associated with land ownership. The share of rented farmland varies widely by region and production activity. However, for some farms it constitutes an important component of farm operating expenses. Since 2002, about 38% of agricultural land used in U.S. farming operations has been rented.[22] Some farmland is rented from other farm

[22] ERS, "Land Use, Land Values and Tenure: Farmland Ownership and Tenure," http://www.ers.usda.gov/topics/farmeconomy/land-use,-land-value-tenure/farmland-ownership-and-tenure.aspx.

operations—nationally about 8% of all land in farms in 2012 (the most recent year for which data are available)—and thus constitutes a source of income for some operator landlords. However, the majority of rented land in farms is rented from nonoperating landlords. Nationally in 2012, 30% of all land in farms was rented from someone other than a farm operator. Total net rent to nonoperator landlords is projected to be up slightly (+1.3%) at $15.7 billion in 2018. Average cash rental rates for 2018 are up slightly year-over-year ($138 per acre versus $136 in 2017). Although rental rates—which were set the preceding fall of 2017 or in early spring of 2018—dipped in 2016, they still reflect the high crop prices and large net returns of the preceding several years (especially the 2011-2014 period) and have yet to decline to pre-2011 levels (Figure 19). The national rental rate for crop land peaked at $144 per acre in 2015. It is estimated at $138 per acre in 2018.[23]

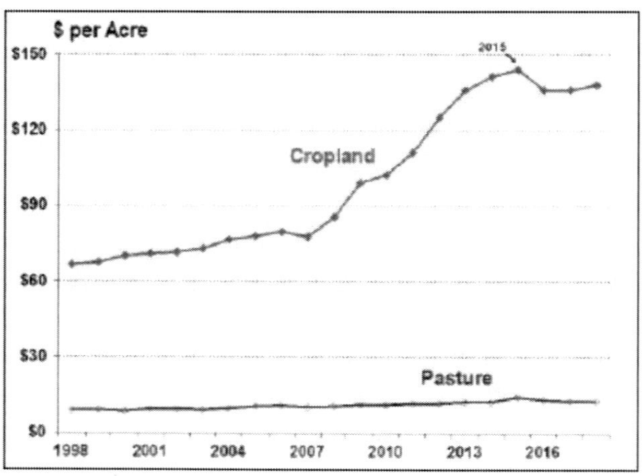

Source: ERS, "2018 Farm Income Forecast," August 30, 2018. All values are nominal. Amounts for 2018 are forecasts.

Figure 19. NASS. *Agricutlrual Land Values*, August 2, 2018. All values are nominal.

[23] Local and regional land rental rates may vary substantially from the national average.

Agricultural Trade Outlook

U.S. agricultural exports have been a major contributor to farm income, especially since 2005. As a result, the downturn in those exports that followed 2014's peak of $152.3 billion both coincided with and deepened the downturn in farm income that started in FY2015. In general, since reaching $137.4 billion in FY2011, U.S. agricultural exports have fluctuated between $130 billion and $152 billion (Figure 20).

Key U.S. Agricultural Trade Highlights

- USDA projects U.S. agricultural exports at $140 billion in FY2018 (unchanged from 2017's total). This includes processed and unprocessed agricultural product exports.
- USDA also projects that U.S. agricultural imports will be lower at $117.0 billion (-2%), with a resulting agricultural trade surplus of $23.0 billion (+8%).
- A substantial portion of the increase in U.S. agricultural exports since 2010 has been due to higher-priced grain and feed shipments, record oilseed exports to China, and growing animal product exports to East Asia.
- The top three markets for U.S. agricultural exports are China, Canada, and Mexico in that order. Together, these three countries have accounted for 46% of total U.S. agricultural exports during the five-year period FY2014-FY2018 (Figure 21). However, their combined share is projected down to 44% in FY2019.
- The fourth- and fifth-largest U.S. export markets are the European Union and Japan, which have accounted for a combined 17% of U.S. agricultural exports during the FY2014-FY2018 period. This same share is projected to continue in FY2019 (Figure 21). Thus, these two markets have shown relatively limited growth in recent years when compared with the rest of the world.

- The "Rest of World" (ROW) component of U.S. agricultural trade—South and Central America, the Middle East, Africa, and Southeast Asia—has shown strong import growth in recent years. ROW is expected to account for 39% of U.S. agricultural exports in FY2019.

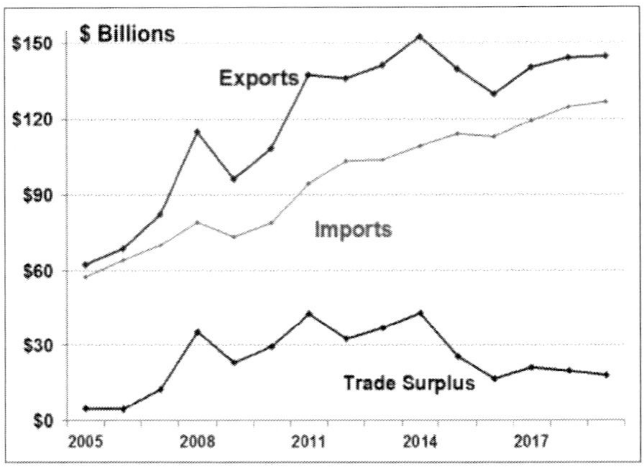

Source: ERS, Outlook for U.S. Agricultural Trade, AES-105, August 29, 2018. Amounts for 2018 and 2019 are projected.

Figure 20. U.S.Agricultural Trade Since 2005, Nominal Values.

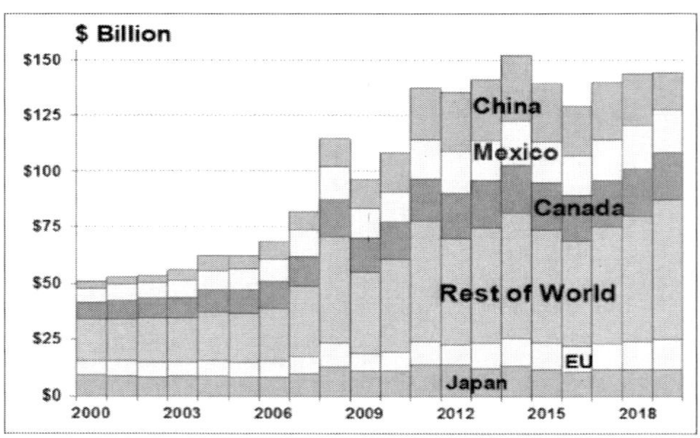

Source: ERS, Outlook for U.S. Agricultural Trade, AES-105, August 29, 2018. Amounts for 2018 and 2019 are projected.

Note: Data are for fiscal years (October to September).

Figure 21. U.S. Agricultural Exports Have Leveled Off Since FY2011.

- Over the past four decades, U.S. agricultural exports have experienced fairly steady growth in shipments of high-value products (Figure 22). As grain and oilseed prices decline, so will the bulk value share of U.S. exports.
- Bulk commodity shipments (primarily wheat, rice, feed grains, soybeans, cotton, and unmanufactured tobacco) are forecast at a 34.5% share of total U.S. agricultural exports in FY2019 at $49.9 billion. This compares with an average share of over 60% during the 1970s and 1980s.
- In contrast, high-valued export products—including horticultural products, livestock, poultry, and dairy—are forecast at $94.6 billion for a 65.5% share of U.S. agricultural exports in FY2019.

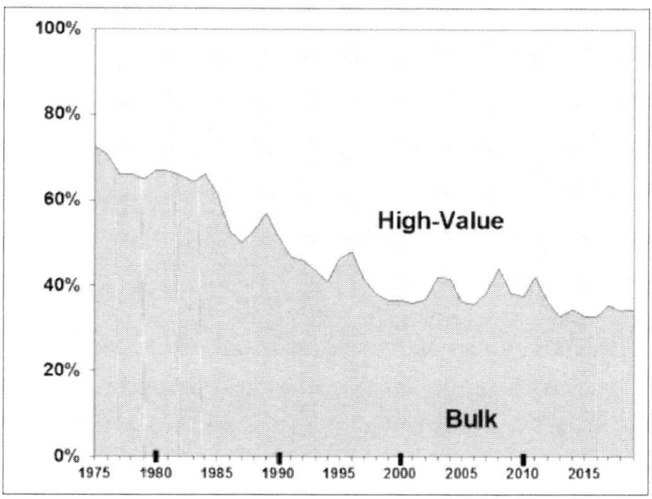

Source: ERS, Outlook for U.S. Agricultural Trade, AES-105, August 29, 2018. Amounts for 2018 and 2019 are projected.

Figure 22. U.S. Agricultural Trade: Bulk vs. High-Value Shares.

The share of agricultural production (based on value) sold outside the country indicates the level of U.S. agriculture's dependence on foreign markets, as well as the overall market for U.S. agricultural products.

The U.S. Export Share Measurement for Agriculture

Because agricultural and food exports consist of farm commodities and their manufactured products, a substantial component of export value represents value-added from marketing and processing. This value-added must be accounted for in measuring the U.S. export share. With this in mind, ERS calculates the export value share for agriculture as follows.[24] The numerator includes aggregated export values for all agricultural products—including bulk commodities and manufactured products. The denominator includes the total value of U.S. farm and manufactured agricultural production—estimated as the sum of farm cash receipts for crop and livestock production plus the value added by agricultural manufacturers. The value-added amount for agricultural processing is from the U.S. Census Bureau's Annual Survey of Manufacturers.

As a share of total farm and manufactured agricultural production, U.S. exports are estimated to account for 19.8% of the overall market for agricultural products from 2008 through 2016—the most recent data year for this calculation (Figure 23). The export share of agricultural production varies significantly by product category.

- At the upper end of the range for export shares, the bulk food grain export share has varied between 50% and 80% since 2008, while the oilseed export share has ranged between 47% and 58%.
- Mid-spectrum range of export shares include the export share for fruit and tree nuts, which has ranged from 37% to 45%, while meat products have ranged from 27% to 41%.
- At the low end of the spectrum, the export share of vegetable and melon sales has ranged from 15% to 18%, the dairy products export share from 9% to 24%, and the agricultural-based beverage export share between 7% and 13%.

[24] The methodology is described at ERS, "U.S. Export Share of Production," https://www.ers.usda.gov/topics/international-markets-us-trade/us-agricultural-trade/.

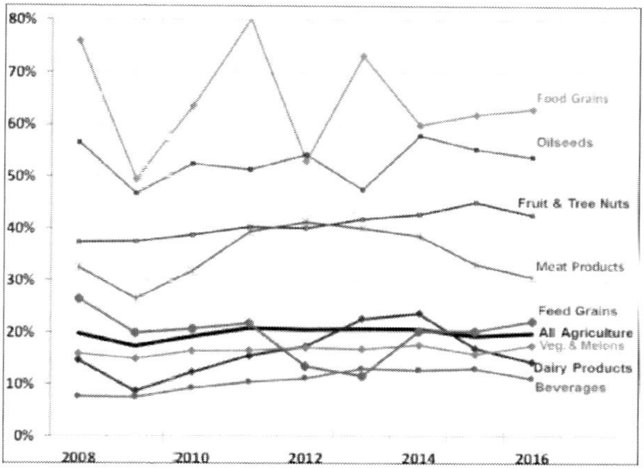

Source: ERS, "U.S. Export Share of Production," as of February 9, 2018.

Note: The export value share includes aggregated export values for all agricultural products—including bulk commodities and manufactured products—in the numerator and the total value of U.S. farm and manufactured agricultural production—estimated as the sum of farm cash receipts for crop and livestock production plus the value added by agricultural manufacturers—in the denominator.

Figure 23. U.S. Export Share of Farm and Manufactured Agricultural Production.

FARM ASSET VALUES AND DEBT

The U.S. farm income and asset-value situation and outlook suggest a relatively stable financial position heading into 2018 for the agriculture sector as a whole—but with considerable uncertainty regarding the downward outlook for prices and market conditions for the sector and an increasing dependency on international markets to absorb domestic surpluses:

- Farm asset values—which reflect farm investors' and lenders' expectations about long-term profitability of farm sector investments—are projected to be up 1.6% in 2018 to a nominal $3 trillion (Table A-3). In inflation-adjusted terms (using 2017 dollars), farm asset values peaked in 2014 (Figure 24).

- Nominally higher farm asset values are expected in 2018 due to strength in real estate values (+2.1%) as well as a slight increase in non-real estate values (+0.6%). Real estate traditionally accounts for the bulk of the total value of farm sector assets—nearly an 81% share.
- Crop land values are closely linked to commodity prices. The leveling off of crop land values in 2017 reflects mixed forecasts for commodity prices (corn, soybeans, and cotton lower; wheat, rice, and livestock products higher) and the uncertainty associated with international commodity markets (Figure 25).
- Total farm debt is forecast to rise to $406.9 billion in 2018 (+3.5%) (Table A-3).
- Farm equity (or net worth, defined as asset value minus debt) is projected to be up slightly (+0.8%) at $2.6 trillion in 2018 (Table A-3).
- The farm debt-to-asset ratio is forecast up in 2018 at 13.4%, the highest level since 2003 but still relatively low by historical standards (Figure 26).

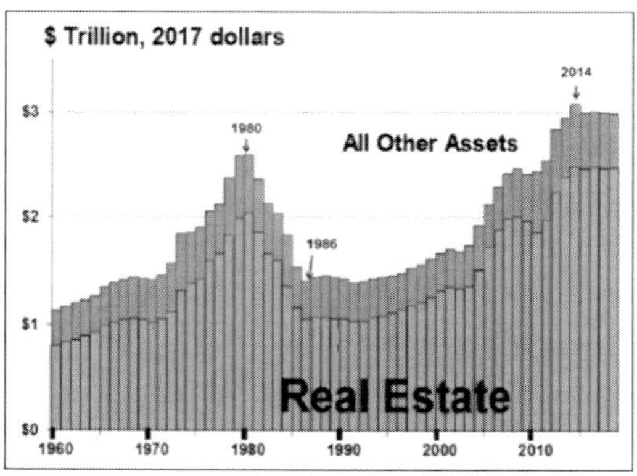

Source: ERS, "2018 Farm Income Forecast," August 30, 2018. All values are adjusted for inflation using the chain-type GDP deflator, 2017 = 100, OMB, Historical Tables, Table 10.1; 2018 and 2019 are forecasts.

Note: Non-real estate assets include financial assets, inventories of agricultural products, and the value of machinery and motor vehicles.

Figure 24. Real Estate Assets Comprise 81% of Total Farm Sector Assets in 2018.

Measuring Farm Wealth: The Debt-to-Asset Ratio

A useful measure of the farm sector's financial wherewithal is farm sector net worth as measured by farm assets minus farm debt. A summary statistic that captures this relationship is the debt-to-asset ratio.

Farm assets include both physical and financial farm assets. *Physical assets* include land, buildings, farm equipment, on-farm inventories of crops and livestock, and other miscellaneous farm assets. *Financial assets* include cash, bank accounts, and investments such as stocks and bonds.

Farm debt includes both business and consumer debt linked to real estate and non-real estate assets (e.g., financial assets, inventories of agricultural products, and the value of machinery and motor vehicles) of the farm sector.

The *debt-to-asset ratio* compares the farm sector's outstanding debt related to farm operations relative to the value of the sector's aggregate assets. Change in the debt-to-asset ratio is a critical barometer of the farm sector's financial performance, with lower values indicating greater financial resiliency. A lower debt-to-asset ratio suggests that the sector is better able to withstand short-term increases in debt related to interest rate fluctuations or changes in the revenue stream related to lower output prices, higher input prices, or production shortfalls.

The largest single component in a typical farmer's investment portfolio is farmland. As a result, real estate values affect the financial well-being of agricultural producers and serve as the principal source of collateral for farm loans.

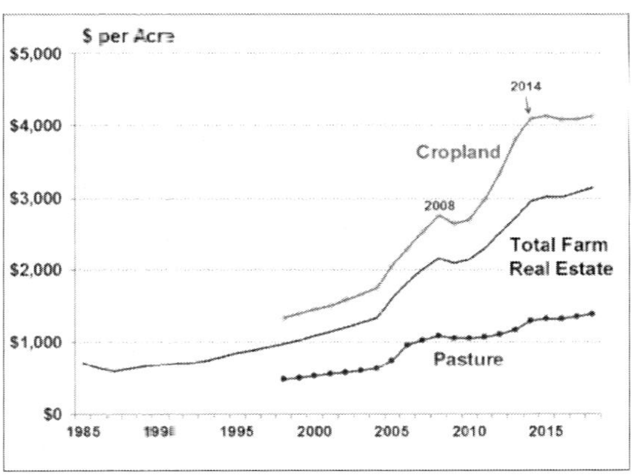

Source: NASS, Land Values 2018 Summary, August 2018.

Notes: Farm real estate value measures the value of all land and buildings on farms. Cropland and pasture values are available only since 1998. All values are nominal.

Figure 25. U.S. Average Farm Land Values, 1985-2018.

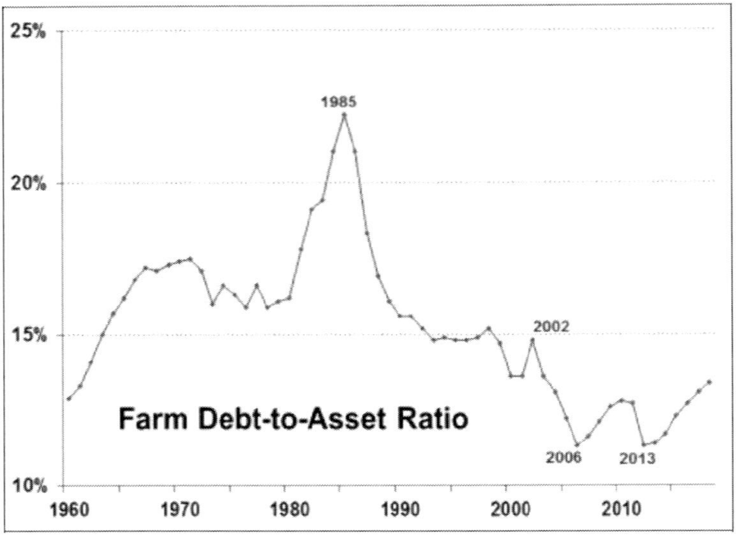

Source: ERS, "2018 Farm Income Forecast," August 30, 2018. Values for 2018 are forecasts.

Figure 26. U.S. Farm Debt-to-Asset Ratio, 1960-2018.

AVERAGE FARM HOUSEHOLD INCOME

Farm household wealth is derived from a variety of sources.[25] A farm can have both an on-farm and an off-farm component to its balance sheet of assets and debt. Thus, the well-being of farm operator households is not equivalent to the financial performance of the farm sector or of farm businesses because of the inclusion of other stakeholders in farming—including landlords and contractors—and because farm operator households often have nonfarm investments, jobs, and other links to the nonfarm economy.

[25] ERS, "Farm Household Well-Being," http://www.ers.usda.gov/topics/farm-economy/farm-household-wellbeing.aspx.

On-Farm vs. Off-Farm Income Shares

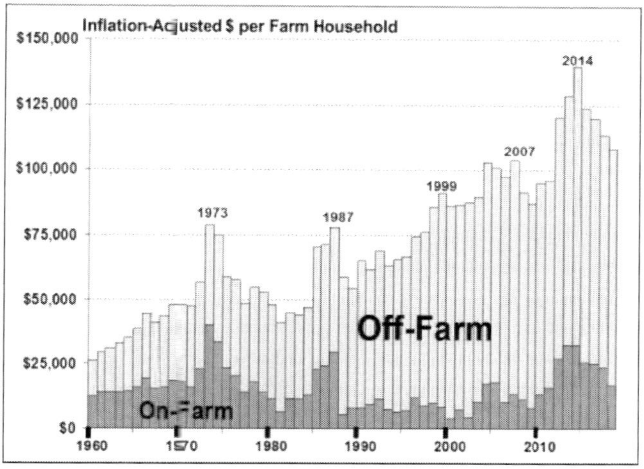

Source: ERS, "2018 Farm Income Forecast," August 30, 2018. All values are adjusted for inflation using the chain-type GDP deflator, 2017 = 100, OMB, Historical Tables, Table 10.1. Values for 2018 are forecasts.

Figure 27. U.S.Average Farm Household Income, by Source, 1960-2018.

Average farm household income (sum of on- and off-farm income) is projected at $109,786 in 2018 (*Table A-2*), down 8.2% from 2017 and well below the record of $134,164 in 2014. About 16% ($17,119) of total farm household income is from farm production activities, and the remaining 84% ($92,667) is earned off the farm (including financial investments). The share of farm income derived from off-farm sources had increased steadily for decades but peaked at about 95% in 2000 (*Figure 27*).

U.S. Total vs. Farm Household Average Income

- Since the late 1990s, farm household incomes have surged ahead of average U.S. household incomes (Figure 28 and Figure 29).

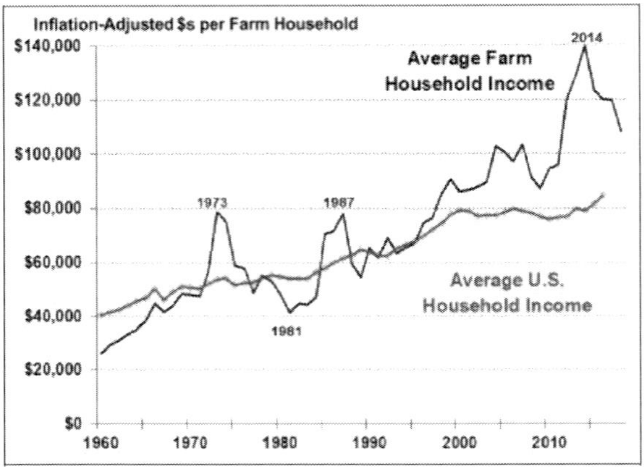

Source: ERS, "2018 Farm Income Forecast," August 30, 2018. All values are adjusted for inflation using the chain-type GDP deflator, 2017 = 100, OMB, Historical Tables, Table 10.1. Values for 2018 are forecasts.

Figure 28. Farm Household Income Has Been Above U.S.Average, 1960-2018.

- In 2016 (the last year for which comparable data were available), the average farm household income of $117,918 was about 42% higher than the average U.S. household income of $83,143 (Table A-2).

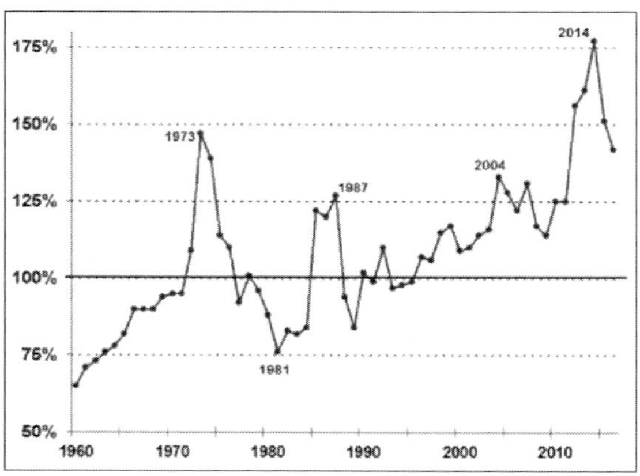

Source: ERS, "2018 Farm Income Forecast," August 30, 2018; 2016 is the last year with comparable data.

Figure 29. Ratio of Average U.S. Farm to National Household Income.

APPENDIX. SUPPORTING CHARTS AND TABLES

USDA Monthly Farm Prices Received Charts

The following set of four charts (*Figure A-1* to *Figure A-4*) presents USDA data on monthly farm prices received for several major farm commodities—corn, soybeans, wheat, upland cotton, rice, milk, cattle, hogs, and chickens. The data is presented in an indexed format where monthly price data for year 2010 = 100 to facilitate comparisons.

USDA Farm Income Data Tables

Three tables at the end of this chapter (*Table A-1* to *Table A-3*) present aggregate farm income variables that summarize the financial situation of U.S. agriculture. In addition, *Table A-4* presents the annual average farm price received for several major commodities, including the USDA forecast for the 2017-2018 and 2018-2019 marketing years.

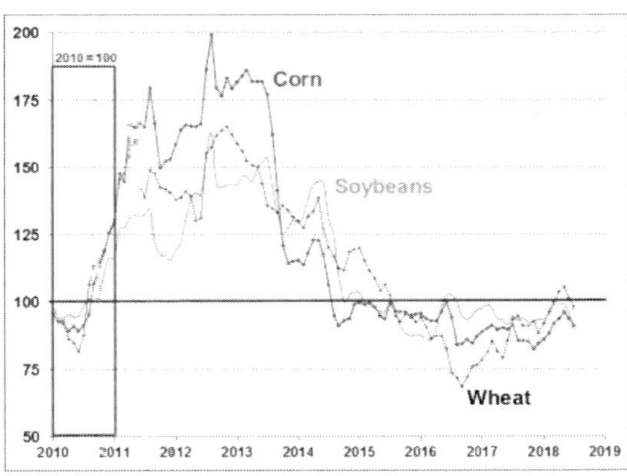

Source: NASS, Agricultural Prices, August 29, 2018. Calculations by CRS.
Notes: Prices are indexed to 2010 = 100 to permit relative comparisons.

Figure A-1. Monthly Farm Prices for Corn, Soybeans, and Wheat, Indexed Dollars.

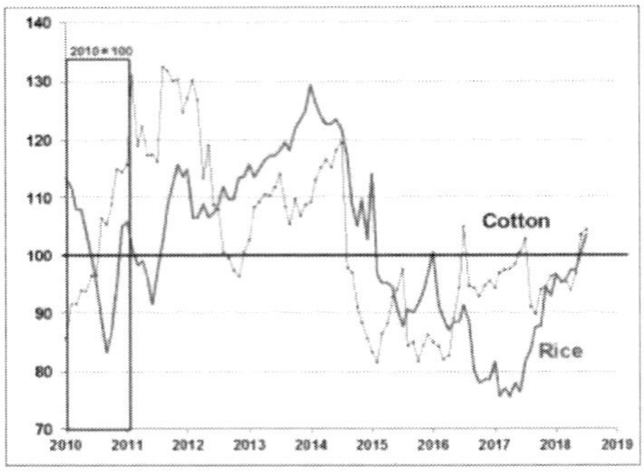

Source: NASS, Agricultural Prices, August 29, 2018. Calculations by CRS.

Figure A-2. Monthly Farm Prices for Cotton and Rice, Indexed Dollars.

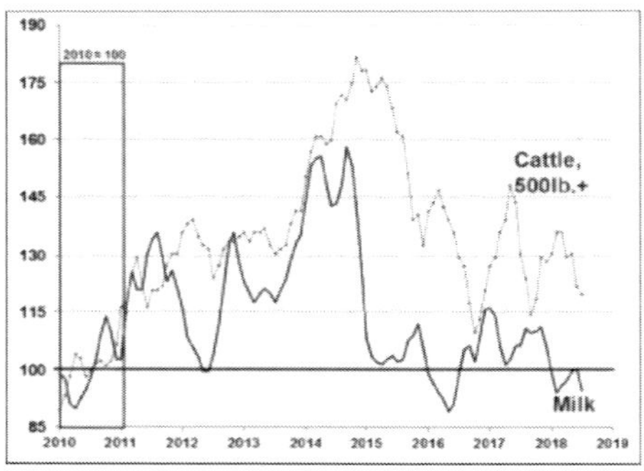

Source: NASS, Agricultural Prices August 29, 2018. Calculations by CRS.
Notes: Prices are indexed to 2010 = 100 to permit relative comparisons.

Figure A-3. Monthly Farm Prices for All-Milk and Cattle (500+ lbs.), Indexed Dollars.

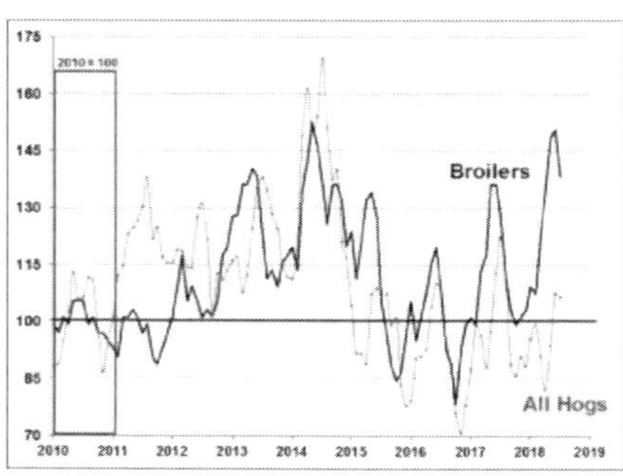

Source: NASS, Agricultural Prices, August 29, 2018. Calculations by CRS.
Notes: Prices are indexed to 2010 = 100 to permit relative comparisons.

Figure A-4. Monthly Farm Prices for All Hogs and Broilers, Indexed Dollars.

Table A-1. Annual U.S. Farm Income Since 2011 ($ billions)

Item	2011	2012	2013	2014	2015	2016	2017	2018a	Change (%)a
1.Cash receipts	365.9	401.4	403.6	424.2	377.1	357.9	374.3	374.0	-0.1%
Crops[b]	201.0	231.6	220.8	211.4	187.6	195.0	198.3	197.8	-0.3%
Livestock	164.8	169.8	182.7	212.8	189.5	162.8	176.0	176.2	0.1%
2. Government payments[c]	10.4	10.6	11.0	9.8	10.8	13.0	11.5	9.5	-18.6%
Fixed direct payments[d]	4.7	4.7	4.3	0.5	0.0	0.0	0.0	0.0	NA
CCP-PLC-ARC[e]	0.0	0.0	0.0	0.0	5.3	8.2	7.0	3.5	-50.3%
Marketing loan benefits[f]	0.0	0.0	0.0	0.1	0.2	0.2	0.0	0.0	NA
Conservation	3.7	3.7	3.7	3.6	3.6	3.8	3.8	3.9	2.9%
Ad hoc and emergency[g]	1.3	1.1	2.1	5.0	1.8	0.7	0.7	1.7	143.6%
All other[h]	0.7	1.1	0.9	0.7	0.0	0.4	0.0	0.4	NA
3. Farm-related income[i]	30.8	39.2	41.0	36.6	34.4	27.9	32.9	34.7	5.6%
4. Gross cash income (1+2+3)	407.0	451.3	455.5	470.6	422.3	398.7	418.7	418.2	-0.1%
5. Cash expenses[j]	283.6	316.1	320.0	339.0	315.9	303.8	314.8	326.7	3.8%
6. NET CASH INCOME	123.2	135.3	135.6	131.6	106.3	94.9	104.0	91.5	-12.0%
7. Total gross revenues[k]	420.4	449.8	483.8	483.1	440.5	411.8	429.6	431.6	0.5%
8. Total production expenses[l]	306.9	353.3	360.1	390.7	359.0	350.1	354.1	365.9	3.3%
9. NET FARM INCOME	113.6	96.5	123.8	92.4	81.5	61.6	75.5	65.7	-13.0%

Source: ERS, Farm Income and Wealth Statistics; U.S. and State Farm Income and Wealth Statistics, updated as of August 30, 2018. NA = not applicable.

[a]Data for 2018 are USDA forecasts. Change represents year-to-year projected change between 2017 and 2018.

[b]Includes Commodity Credit Corporation loans under the farm commodity support program.

[c] Government payments reflect payments made directly to all recipients in the farm sector, including landlords. The nonoperator landlords' share is offset by its inclusion in rental expenses paid to these landlords and thus is not reflected in net farm income or net cash income.

[d] Direct payments include production flexibility payments of the 1996 Farm Act through 2001 and fixed direct payments under the 2002 Farm Act since 2002.

[e]CCP = counter-cyclical payments. PLC = Price Loss Coverage. ARC = Agricultural Risk Coverage.
[f]Includes loan deficiency payments, marketing loan gains, and commodity certificate exchange gains.
[g]Includes payments made under the ACRE program, which was eliminated by the 2014 farm bill (P.L. 113-79).
[h]Cotton ginning cost-share, biomass crop assistance program, milk income loss, tobacco transition, and other miscellaneous program payments.
[i]Income from custom work, machine hire, agri-tourism, forest product sales, and other farm sources.
[j]Excludes depreciation and perquisites to hired labor.
[k]Gross cash income plus inventory adjustments, the value of home consumption, and the imputed rental value of operator dwellings.
[l]Cash expenses plus depreciation and perquisites to hired labor.

Table A-2. Average Annual Income per U.S. Household, Farm Versus All, 2011-2018 ($ per household)

	2011	2012	2013	2014	2015	2016	2017	2018
Average U.S. farm income by source								
On-farm income	$14,625	$25,038	$30,639	$31,025	$24,740	$24,731	$23,678	$17,119
Off-farm income	$72,665	$86,486	$90,481	$103,140	$95,140	$93,187	$89,817	$92,667
Total farm income	$87,290	$111,524	$121,120	$134,165	$119,880	$117,918	$119,598	$109,786
Average U.S. household income	$69,677	$71,274	$75,195	$75,738	$79,263	$83,143	NA	NA
Farm household income as share of U.S. avg. household income (%)	125%	156%	161%	177%	151%	142%	NA	NA

Source: ERS, Farm Household Income and Characteristics, principal farm operator household finances, data set updated as of August 30, 2018, http://www.ers.usda.gov/data-products/farm-household-income-and-characteristics.aspx.

Note: NA = not available. Data for 2017 and 2018 are USDA forecasts.

Table A-3. Average Annual Farm Sector Debt-to-Asset Ratio, 2011-2018 ($ billions)

	2011	2012	2013	2014	2015	2016	2017	2018
Farm assets	2,318.7	2,638.2	2,776.1	2,949.2	2,909.7	2,956.5	2993.1	3,028.6
Farm debt	294.5	297.5	315.3	345.2	356.7	374.2	393.0	406.9
Farm equity	2,024.2	2,340.7	2,460.8	2,604.0	2,552.9	2,582.3	2,600.0	2,621.8
Debt-to-asset ratio (%)	*12.7%*	*11.3%*	*11.4%*	*11.7%*	*12.3%*	*12.7%*	*13.1%*	*13.4%*

Source: ERS, Farm Income and Wealth Statistics; U.S. and State Farm Income and Wealth Statistics, updated as of August 30, 2018, http://www.ers.usda.gov/data-products/ farm-income-and-wealth-statistics.aspx.

Note: Data for 2017 and 2018 are USDA forecasts.

Table A-4. U.S. Farm Prices and Support Rates for Selected Commodities Since 2013-2014 Marketing Year

Commodity[a]	Unit	Year	2013-2014	2014-2015	2015-2016	2016-2017	2017-2018	2018-2019[b]	% Chg[c] 2017 to 2018	2019-2020[b]	% Chg[d] 2018 to 2019	Loan Rate[e]	Reference Price
Wheat	$/bu	Jun-May	6.87	5.99	4.89	3.89	4.72	4.70-5.50	8.1%	—	—	2.94	5.50
Corn	$/bu	Sep-Aug	4.46	3.70	3.61	3.36	3.40	3.00-4.00	2.9%	—	—	1.95	3.70
Sorghum	$/bu	Sep-Aug	4.28	4.03	3.31	2.79	3.25	2.80-3.80	1.5%	—	—	1.95	3.95
Barley	$/bu	Jun-May	6.06	5.30	5.52	4.96	4.47	4.10-5.30	5.1%	—	—	1.95	4.95
Oats	$/bu	Jun-May	3.75	3.21	2.12	2.06	2.59	2.50-3.10	8.1%	—	—	1.39	2.40
Rice	$/cwt	Aug-Jul	16.30	13.40	12.10	10.40	12.60	11.20-12.20	-7.1%	—	—	6.50	14.00
Soybeans	$/bu	Sep-Aug	13.00	10.10	8.95	9.47	9.35	7.35-9.85	-6.4%	—	—	5.00	8.40
Soybean Oil	¢/lb	Oct-Sep	38.23	31.60	29.86	32.48	30.0	28.00-32.00	0.0%	—	—	—	—
Soybean Meal	$/st	Oct-Sep	489.94	368.49	324.6	316.9	345	290-330	-10.1%	—	—	—	—
Cotton, Upland	¢/lb	Aug-Jul	77.9	61.3	61.2	68	68	70-80	10.3%	—	—	45-52	none
Choice Steers	$/cwt	Jan-Dec	125.89	154.56	148.12	120.86	121.52	115-117	-4.5%	113-122	1.3%	—	—

Table A-4. (Continued)

Commodity[a]	Unit	Year	2013-2014	2014-2015	2015-2016	2016-2017	2017-2018	2018-2019[b]	% Chg[c] 2017 to 2018	2019-2020[b]	% Chg[d] 2018 to 2019	Loan Rate[e]	Reference Price
Barrows/Gilts	$/cwt	Jan-Dec	64.05	76.03	50.23	46.16	50.48	43-44	-13.8%	39-42	-9.0%	—	—
Broilers	¢/lb	Jan-Dec	99.7	104.90	90.5	84.3	93.5	96-98	3.7%	92-100	-1.0%	—	—
Eggs	¢/doz	Jan-Dec	124.7	142.3	181.8	85.7	100.9	140-142	39.7%	119-129	-12.1%	—	—
Milk	$/cwt	Jan-Dec	20.05	23.97	17.12	16.30	17.65	16.30-16.50	-7.1%	16.75-17.75	5.2%	—	—

Source: Various USDA agency sources as described in the notes below. bu = bushels, cwt = 100 pounds, lb = pound, st = short ton (2,000 pounds), doz = dozen.

[a] Price for grains and oilseeds are from USDA, World Agricultural Supply and Demand Estimates (WASDE), September 12, 2018. Calendar year data are for the first year. For example, 2018-2019 = 2018.—= no value, and USDA's out-year 2019-2020 crop price forecasts will first appear in the May 2019 WASDE. Soybean and livestock product prices are from USDA, Agricultural Marketing Service: soybean oil—Decatur, IL, cash price, simple average crude; soybean meal—Decatur, IL, cash price, simple average 48% protein; choice steers—Nebraska, direct 1,100-1,300 lbs.; barrows/gilts—national base, live equivalent 51%-52% lean; broilers—wholesale, 12- city average; eggs—Grade A, New York, volume buyers; and milk—simple average of prices received by farmers for all milk.

[b] Data for 2018-2019 are USDA forecasts. Data for 2019-2020 are USDA projections.

[c] Percent change from 2016-2017, calculated using the difference from the midpoint of the range for 2017-2018 with the estimate for 2016-2017.

[d] Percent change from 2017-2018, calculated using the difference from the midpoint of the range for 2018-2019 with the estimate for 2017-2018.

[e] Loan rate and reference prices are for the 2017-2018 market year. See CRS Report R43076, The 2014 Farm Bill (P.L. 113-79): Summary and Side-by-Side.

In: Agricultural Issues
Editor: Michael Butlin

ISBN: 978-1-53615-474-0

Chapter 7

THE 2014 FARM ACT AGRICULTURE RISK COVERAGE, PRICE LOSS COVERAGE, AND SUPPLEMENTAL COVERAGE OPTION PROGRAMS' EFFECTS ON CROP REVENUE*

Erik J. O'Donoghue, Ashley E. Hungerford, Joseph C. Cooper, Thomas Worth and Mark Ash

SUMMARY

The 2014 Farm Act provides eligible U.S. farmers with new commodity supports in the Agriculture Risk Coverage (ARC), the Price Loss Coverage (PLC), and the Supplemental Coverage Option (SCO) programs. These programs help producers with revenue losses generally not covered by traditional crop insurance policies. Interactions, both among these programs and between these programs and the Federal Crop Insurance (FCI) program, determine the nature and magnitude of support available to producers. This chapter provides an analysis of these

* This is an edited, reformatted and augmented version of United States Department of Agriculture Economic Research Report, Publication No. 204, dated January 2016.

programs with a focus on how various combinations of the programs impact producer revenue and its variability, producer well-being, and expected program costs. The report finds that these programs' effectiveness are influenced by historical prices, expected prices, and FCI coverage rates. High historic crop prices combined with low expected prices since the enactment of the 2014 Farm Act led to higher enrollment of producers in the ARC program in 2015 relative to that in the SCO program.

Keywords: 2014 Farm Act, Agriculture Risk Coverage, crop insurance, crop revenue, Supplemental Coverage Option

What Is the Issue?

The structure of U.S. agricultural commodity support changed significantly under the 2014 Farm Act. The new programs—in particular, Price Loss Coverage (PLC), Supplemental Coverage Option (SCO), and Agriculture Risk Coverage (ARC)—have linkages with each other and with the pre-existing Federal Crop Insurance (FCI) program. (In this chapter, the term "ARC" refers to the most popular version of ARC (ARC-County), which is an area-based, rather than individual-based, commodity program.) While PLC builds on the old Countercyclical Payment program, SCO and ARC are known as "shallow loss" programs, covering losses typically not covered by the "deep loss" FCI program. Understanding these new programs and how the available combinations of programs can affect crop revenue provides information on agricultural producers' enrollment decisions, the programs' impact on producers' risk and revenues, and expected program costs.

What Did the Study Find?

Interactions, both among these programs and the Federal Crop Insurance (FCI) program are complex. The mandatory decision producers had to make to elect either the Agriculture Risk Coverage or Price Loss Coverage programs will last for the duration of the Farm Act and has implications for how they can use crop insurance. For example, if a producer elects ARC, the producer cannot use SCO, a crop insurance policy. If the producer instead elects PLC, the producer can enroll in SCO, which then takes on the traits of the (required) underlying policy it supplements—which can have implications for the type and coverage level of the underlying crop insurance policy that a producer chooses to enroll in.

At first glance, the two major "shallow loss" programs for field crops, ARC and SCO, appear similar. However, like the benefits from the Direct and Countercyclical Program that was repealed with the 2014 Farm Act, ARC payments are not influenced by current production. In contrast, the size of the SCO payments are linked to the expected crop production of the farm for the current year.

Moreover, the ARC program has a "memory" for prices—it relies on historic prices to calculate the potential benefits for the producer. In contrast, the SCO guarantee depends on the higher of the expected prices at planting time (also known as the futures price) or the realized harvest time price. Therefore, while ARC provides benefits that depend on past outcomes, which helps to smooth payments over time, SCO provides intra-year benefits, comparing expected returns at planting time with the actual returns realized at harvest time.

Assuming the futures prices are "close" to the eventual realized prices, these expected prices also matter for the benefits a producer receives from these programs. ERS research results suggest that in an environment with lower expected prices, the Agriculture Risk Coverage program helps to minimize the largest potential losses the most (in other words, ARC helps to increase the lower bound of expected revenues more than SCO would in an environment with low expected prices). In an environment with higher expected prices, the reverse is true and the Supplemental Coverage Option policy helps to minimize a producer's largest potential losses the most.

When producers had to make their decision to elect either ARC or PLC, historic commodity prices had been high while expected commodity prices for the upcoming crop year were low. Likely due to the differences in how the Agriculture Risk Coverage and Supplemental Coverage Option programs work with respect to both high historic prices and low expected commodity prices, producers overwhelmingly elected ARC instead of the Price Loss Coverage program (and hence over SCO). According to USDA's Farm Service Agency, producers elected ARC for 91 percent of corn base acres and 96 percent of soybean base acres. While almost all corn and soybean producers elected ARC, more than one in three wheat base acres were covered with PLC—and hence were eligible for enrollment in the SCO program. Compared to ARC, SCO appears to provide higher benefits for winter wheat, providing slightly higher average revenues while generating similar potential low-end losses as the ARC program, which could help explain why a significant portion of wheat producers made different choices than corn and soybean producers.

HOW WAS THE STUDY CONDUCTED?

The analysis translates 2014 Farm Act terms into quantitative functions. Because no data exist for these programs—since they have only recently been enacted into law—ERS researchers used a model to simulate revenue outcomes for a representative (typical) producer for each county that produced corn, soybeans, and wheat. This model was used to generate distributions of simulated crop prices and yields, centered on their expected values at planting time in 2014.

For each crop, nonparametric county-level yield distributions are generated for each county for which USDA's National Agricultural Statistics Service (NASS) has reported data each year from 1975 through 2013. This amounted to 1,001 counties for corn, 889 counties for soybeans, and 510 counties for wheat. The price distribution is also generated non-parametrically, based on planting time and harvest time futures prices over the same period.

The analysis maintains the historical correlations of yields across all counties and between the county yields and prices using an empirical approach that helps describe the historic relationships between the two variables. The yield distribution for a typical (representative) farmer in each county is generated by inflating the county-level yield variability based on farm yield information implicit in actual crop insurance premium rates for each county.

The model makes 10,000 draws from each county and farm yield distribution as well as from the price distribution. Payments, net revenue, and total revenue (net revenue plus the payments) are then calculated for each of the 10,000 price and yield pairs to generate distributions for each of these variables, providing researchers with the data used in the analysis.

INTRODUCTION

Crop revenue is subject to both price risk and yield risk. Price risk is a systemic risk in the sense that it is correlated among farmers (e.g., Miranda and Glauber, 1997; Cooper, 2009; USDA/ERS, 2014a). Since price is determined by domestic production, stocks, and trade, as well as demand factors, no one farmer controls price. Therefore, all farmers receive the same price for the same commodity (after adjusting for arbitrage), making price strongly correlated across farmers. Yield risk has both systemic and idiosyncratic components (Miranda, 1991). The main source of systemic risk in yields is weather. Droughts, in particular, have proven to be a major cause of systemic losses, such as the 2012 drought in the Corn Belt (USDA/RMA, 2014d). By definition, idiosyncratic risk is uncorrelated across farmers and has a variety of causes. Furthermore, prices and yields can be correlated at the farm level and can vary across farms while producer financial-risk profiles differ, which can affect producers' willingness to take on price and yield risk. The 2014 Farm Act enacted various programs to help producers reduce the risks they face.

The 2014 Farm Act includes several new farm programs such as the Price Loss Coverage (PLC), the Agriculture Risk Coverage (ARC), and the Supplemental Coverage Option (SCO). The first two, PLC and ARC, replace the repealed Direct Payments (DPs), Counter-cyclical Payments (CCPs), and

Average Crop Revenue Election (ACRE) programs found in the commodities title (Title I) of the 2008 Farm Act. ARC and PLC are administered by the U.S. Department of Agriculture's (USDA) Farm Service Agency (FSA). The SCO policy is a new entrant into the crop insurance title (Title XI), managed by USDA's Risk Management Agency (RMA).

This chapter focuses on three of the new programs—ARC, PLC, and SCO. ARC and SCO are colloquially known as "shallow loss" programs because they address the smaller revenue losses typically not covered under traditional Federal crop insurance policies (policies that cover the "deep losses" potentially felt by producers). Each of these programs affects agricultural risk exposure, but does so in different ways. Moreover, these new programs interact with each other. For example, producers can elect to enroll in ARC or PLC, but not both. If the producer elects ARC, that producer is no longer eligible to enroll in SCO. Further eligibility constraints and interactions exist. However, the empirical differences between the revenue support provided by SCO and ARC remain unexplored.

This study provides a novel graphical approach to provide the intuition behind the mechanics of how payments differ among these programs. In addition, we use a new advanced statistical approach to empirically examine the impacts of these programs on crop revenue, with particular emphasis on the extent to which they can reduce the farmer's downside revenue risk. This empirical analysis examines the impacts of these programs on crop revenue both as standalone programs (even when not available as such) as well as in conjunction with traditional Federal crop insurance. The analysis summarizes the empirical results across all counties for which USDA reports yields and for several individual representative farms from low-, medium-, and high-risk areas of production that allow us to explore differences in the potential ranges of impacts given historic distributions of prices and yields and recent planting time prices and expected yields. The analysis covers the major field crops corn, soybeans, and winter wheat. The results have implications for other eligible crops as well.

2014 FARM ACT PROGRAMS

Several major new programs were introduced in the 2014 Farm Act. After a brief overview of the new programs, we illustrate how they operate, using a county-level example of soybeans.[1]

- ARC provides either individual- or area-based revenue protection without requiring a premium. Payments are paid out to either 85 percent (in the case of area-based) or 65 percent (in the individual-based selection) of base acres. The ARC guarantee is based on an Olympic average (average of the last 5 years, dropping the highest and lowest values) of market-year average prices and yields. Payments begin when realized revenues fall below 86 percent of expected revenues and cover up to 10 percent of the benchmark (expected) revenue.
- SCO provides area-based revenue protection as part of the crop insurance program. Producers pay a premium subsidized at the 65-percent level. The SCO guarantee is based on futures prices and expected yields. SCO supplements the underlying policy in place. SCO payments begin when realized revenues fall below 86 percent of expected revenues and cover up to the level of insurance selected by the producer for the underlying policy.
- PLC provides a price floor for producers, helping mitigate downside price risk.

To minimize overlap between these new programs, growers cannot simultaneously elect ARC and PLC for the same crop on the same farm, nor can a producer elect ARC and enroll in SCO for the same crop and farm. However, electing PLC and enrolling in SCO is allowed. Furthermore, once the election for either ARC or PLC is made, the

[1] The Stacked Income Protection Plan (STAX) is also new in the 2014 Farm Act, but because it is exclusively for upland cotton producers, we do not cover it in this report.

producer is committed to this choice for the duration of the Farm Act (see box, "Producers' Choices").

The payments producers receive depend on prices, yields, and the combination of programs enrolled in. The mechanics behind the various programs is explored, followed by a statistical analysis, to provide a basis for understanding how the payment mechanisms work. The basic payment mechanisms are the same across all covered crops. For the sake of brevity, the numerical discussion in this section focuses on just soybeans. For the numerical illustrations, an average soybean producer from Linn County, IA, is used, assuming that this soybean producer has yields identical to those of the county (since the programs are county-based programs).

Based on actual yields and prices, the expected yield for a typical soybean producer in Linn County, IA, in 2014 is 50 bushels per acre (bu/acre), the Olympic-average yield is also 50 bu/acre, the 5-year Olympic-average price is $12.27 per bushel, and a base insurance price—the average of the daily February (planting month) prices of the November (futures harvest month) Chicago Board of Trade (CBOT) soybean futures contract—is $11.36 per bushel (USDA/RMA, 2014a; USDA/RMA, 2014f).

Agriculture Risk Coverage (ARC)

The ARC program is a revenue-based program in which producers do not need to pay a premium to participate (USDA/FSA, 2014). Similar to the Direct and Countercyclical Program that was repealed in the 2014 Farm Act, the benefits producers receive from the ARC program do not depend on current production outcomes. While producers can enroll in either individual- or county-level coverage, FSA election statistics show that very few participants elected individual-level coverage. Therefore, hereafter in this chapter, the term "ARC" will refer to "ARC-County," the heavily favored area-based version of ARC.[2]

---

[2] [2]Ninety-six percent of soybean, 91 percent of corn, and 66 percent of wheat farmers elected county-level ARC. For more information concerning program election results, see USDA's Farm Service Agency (FSA) website, http://www.fsa.usda.gov/programs-and-

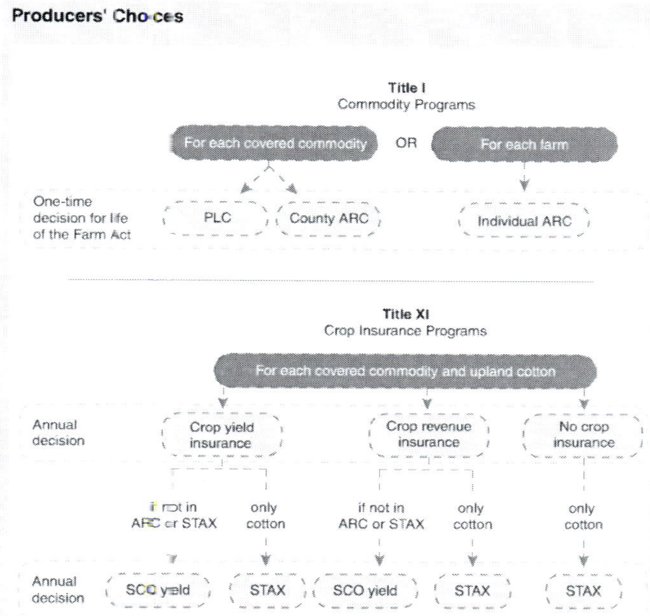

Source: USDA, Economic Research Service.

PLC = Price Loss Coverage, which provides benefits to producers with base acres of wheat, feed grains, rice, oilseeds, peanuts, and pulses (covered commodities) on a commodity-by-commodity basis when market prices drop below the reference price. ARC = Agriculture Risk Coverage, which provides benefits to producers of covered commodities at either the county or individual level. The most popular election, county-based ARC, operates on a commodity-by-commodity basis and provides support when county crop revenue drops below 86 percent of the county benchmark revenue. In the individual ARC case, payments are issued when the actual individual crop revenues, summed across all covered commodities on the farm, are less than the ARC individual guarantee. SCO = Supplemental Coverage Option, which offers producers additional area-based insurance coverage in combination with coverage by traditional crop insurance policies, also beginning with the 2015 crop year. STAX = Stacked Income Protection Plan, which provides revenue insurance policies to producers of upland cotton beginning with the 2015 crop and operates as either a stand-alone policy or in conjunction with an underlying crop insurance policy.

---

services/arcplc_program/. For more information on the specifics of the ARC program, including details of the individual-level ARC program, see the specific link on the USDA/FSA website, http://www.fsa.usda.gov/ Internet/FSA_File/base_acre_reallocate_arc_plc.pdf/.

Producers electing ARC can enroll on a crop-by-crop basis, and the payments they receive are based on 85 percent of their farm's base (historic) acres. The ARC payment calculation centers around the benchmark revenue, a revenue guarantee based on the expected per-acre county revenue. This revenue guarantee can change from year to year because historic national prices and county yields are used to generate the benchmark revenue (USDA/FSA, 2015). For the Linn County, IA, soybean example, the benchmark revenue comes to $613.50 per acre.

The ARC program provides support when area revenue drops below 86 percent of the expected (benchmark) revenue. This support is called the ARC guarantee. The program provides support up to 10 percent of the benchmark (or $61.35 per acre) and is only provided for 85 percent of historic (base) acres. So the maximum level of support for Linn County, IA, comes to roughly $52 per base acre. The per-acre county-ARC payment can be expressed as:

Agriculture Risk Coverage-County (ARC-CO) Per-Acre Payment

ARC guarantee = 0.86*benchmark revenue
Diff = ARC guarantee – realized revenue
Max = 10 percent of benchmark revenue

if Diff ≤ 0	**if Diff > 0 and**	
	if Diff ≤ Max	**if Diff > Max**
then	then	then
Payment = 0	Payment = Diff*0.85	Payment = Max*0.85

Figure 1 shows how ARC payment rates vary for a realized price of $12 per bushel. The maximum level of support ($52 per base acre) is provided up until the realized yield of 38 bushels per acre. After that, support drops until it hits zero at a realized yield of 44 bushels per acre. Given the 86-percent coverage rate and the maximum payment being 10 percent of benchmark revenue, ARC effectively covers (at the area level) realized revenue losses between 76 and 86 percent of the benchmark revenue.

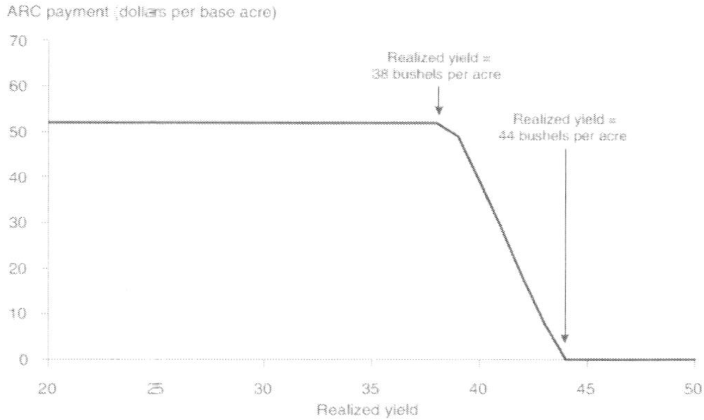

Source: USDA, Economic Research Service.
ARC = Agriculture Risk Coverage program.

Figure 1. ARC payment rates vary by realized yield for a realized price of $12 per bushel.

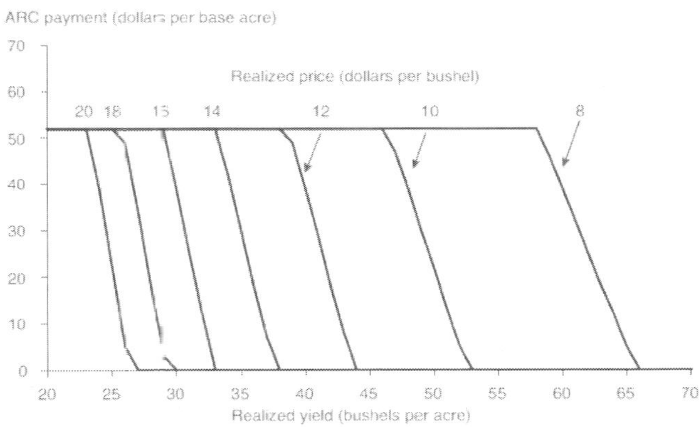

Source: USDA, Economic Research Service.
ARC = Agriculture Risk Coverage program.

Figure 2. ARC payment rates vary as realized prices vary.

Different realized prices generate different payment schedules (fig. 2). Note that for a given expected yield and price, the highest level of support does not change (since it remains constant at 10 percent of the fixed benchmark). As realized prices drop, the maximum level of support is provided for a wider range of yields and vice versa.

Supplemental Coverage Option (SCO)

While ARC is a commodity program, the SCO program works as a crop insurance policy, relying on the underlying individual-level policy to determine the characteristics of support (USDA/RMA, 2014d; USDA/RMA, 2014b). Enrollment requires the farmer to pay a premium, and the SCO policy takes the basic traits of the underlying policy selected by the producer—being either yield or revenue based—and extends it (generally) as a county-based program. For example, if a producer who enrolled in a crop insurance yield protection plan also enrolled in SCO, the producer would receive additional yield protection coverage at the county level. If the producer had a revenue protection policy instead, the SCO would provide additional county-level revenue protection coverage (USDA/ RMA, 2014d).

For this chapter, we assume a revenue-based underlying individual insurance policy with upward price protection (revenue protection, or RP) at a coverage level of 75 percent (the most common policy purchased by producers). With an underlying RP policy, the SCO expected area revenue consists of the higher of the base insurance price and the harvest time price multiplied by the expected yield.[3] With the base insurance price (the 2014 futures price) of $11.36 per bushel and a harvest price below the base insurance price, this comes to an expected per-acre revenue of almost $570. As with the ARC program, area losses between 100 and 86 percent of expected revenue are not covered.

The amount of coverage offered by SCO depends on the coverage level selected for the underlying insurance policy—SCO covers, at the area level, from 86 percent of expected county revenues down to the level of coverage of the underlying insurance policy. For example, if a producer took out an RP policy with a coverage level of 80 percent and supplemented with an SCO policy, the SCO policy would cover—at the area level—from 86 percent of expected revenues down to 80 percent.

[3] The same expected yield is used here as in the ARC example (50 bu/acre). However, it is possible that the expected area yields for SCO and ARC may differ since FSA and RMA use different methodologies to calculate them.

Losses below 80 percent would then be covered by the individual-level RP policy. If, instead, the producer purchased an RP policy with 70-percent coverage supplemented with SCO, the SCO policy would cover from 86 percent down to 70 percent of expected revenues, again at the area level.

Formally, the SCO payment per acre can be expressed as:

$$\text{SCO/acre} = \min\left[\max\left\{\frac{\left(0.86-\frac{Final\ Area\ Rev}{E(Area\ Rev)}\right)}{(0.86-Cov.level)},0\right\},1\right]*E(Farm\ Rev\ per\ acre)*(0.86-Cov.level)-Premium\ per\ acre.$$

where *Rev* denotes "revenue", *E*(.) denotes the expected value of the variable inside the parentheses, and *Cov.* denotes "coverage." The formula in the inner curly brackets is the "revenue payment factor," and to the right of the square brackets, *E*(*Farm Rev per acre*) * (0.86 - *Cov.level*) is the "supplemental protection," to use the RMA terminology.

Similar to ARC payment rates, SCO payment rates vary by the realized price. However, the upward price protection embodied in the underlying RP policy causes the relationship between realized prices, yields, and support rates to differ from those of ARC (fig. 3). The payments in figure 3 are net of the farmer-paid SCO premium, derived from the simulation model. (The model is discussed in more detail later).[4] For simplicity, the figures (but not the simulation analysis in the next section) assume that the expected yield is the same at the farm and area level.

For realized prices below the base insurance price (here \$8/bu and \$10/bu), the SCO payment rates look very similar to those of the ARC program. They both have a maximum level of payment (in our example, \$58 per planted acre) that spans over a wide range of realized yields, which widens as the prices drop. The net payment to the producer is negative (in

[4] We use a simulation model to price the SCO policy. By generating 10,000 draws of revenues, we can apply the SCO program to these draws to estimate payments made to a representative producer for each draw. Averaging over these estimated payments then provides us with an actuarially fair price for the policy, which we use to calculate net returns to the producer in the current section. For more information on the model, see box, "The Empirical Approach," p. 14.

this case, the policy costs $4 per acre) when yields and/or prices are sufficiently high enough that they do not trigger SCO benefits since the producer must pay a premium to obtain SCO coverage.

Because the SCO program takes on the traits of the underlying crop insurance policy (in our example, RP), this means that producers are also insured against a loss of revenue due to price increases. As the realized price exceeds the futures price at the time of planting (in our example, $11.36 per bushel), the SCO payment calculations incorporate the harvest price instead of the futures price. This generates the step-like function seen in figure 3.

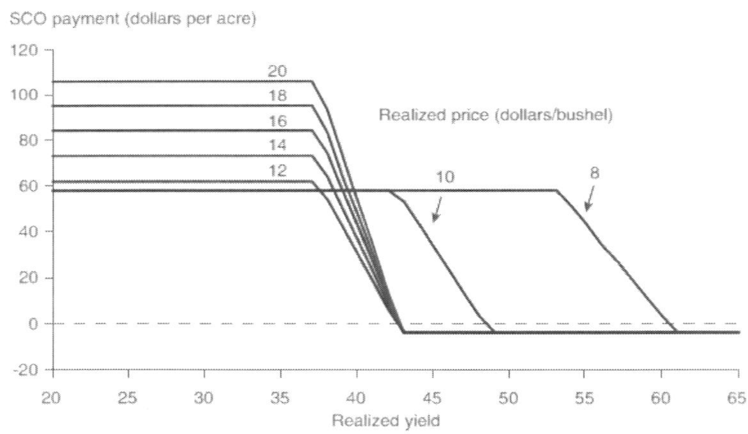

Source: USDA, Economic Research Service.
SCO = Supplemental Coverage Option.

Figure 3. SCO payment rates vary by realized crop price (coverage level on underlying policy = 75 percent).

As noted earlier, unlike ARC, SCO payment rates vary based on the coverage level of the individual underlying revenue policy (fig. 4). For example, with a realized soybean price of $14 per bushel and 65 percent coverage on the underlying policy, a realized yield of 32 bushels per acre would establish the maximum payment rate of $141 per acre. If the individual had an underlying policy with 85-percent coverage, he or she, too, would receive the maximum payment rate, but it would only amount to $6 per acre. Because the producer pays for the SCO policy, for a given price, when yields are high enough that program payments go to zero, the producer will have a different negative return depending on the coverage level chosen.

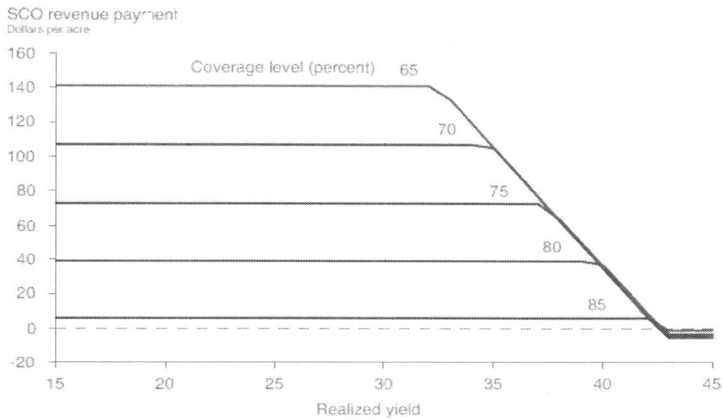

Source: USDA, Economic Research Service.
SCO = Supplemental Coverage Option.

Figure 4. SCO payment rates: the coverage level of the underlying individual policy matters.

Price Loss Coverage

PLC is similar to the recently repealed CCPs, with current reference prices increased from the CCP target prices of the previous Farm Act. The PLC program generates a price floor. Generally, if a program crop's national average market price falls below the statutorily determined reference price for that crop, the producer receives a payment, as expressed here:

PLC/acre =

0.85
* (*Reference price — max[Annual National average market price, Marketing Assistance loan rate]*)
* *payment yield.*

As mentioned earlier, if a producer elects ARC, he or she may not enroll that same acreage in the PLC program (or in the SCO program for that matter). However, if the producer elects to participate in PLC, it is likely that he or she will consider annually purchasing SCO. If the producer does not elect either ARC or PLC, the producer automatically defaults to PLC coverage starting in the 2015 crop year.

ARC versus SCO

ARC and SCO use different benchmark revenues, so the payment levels and ranges of coverage can differ, sometimes dramatically, due to the upside-price protection embodied in SCO when RP is chosen as the underlying policy. Given current conditions and expectations and a 75-percent coverage level on the underlying RP policy, SCO's per-acre payment rates are higher than ARC rates when realized prices are high and realized yields are low, while ARC provides support over a wider range of realized yields. For example, at a realized price of $8 per bushel, ARC will make payments for realized yields up to and including 65 bushels per acre. SCO only provides support for yields up to and including 59 bushels per acre (fig. 5). However, if prices drop far enough, producers begin to receive payments from the PLC program as well, boosting the support they receive. Since the PLC program generates a price floor, if the price falls far enough, producers would receive payments irrespective of the level of yields attained (top-left chart in figure 5).

For higher prices, however, SCO provides higher levels of support over a wider range of yields. In an extreme example, if the harvest price were to hit $20 per bushel, SCO would deliver support up to a maximum of $95 per acre and through yields of 42 bushels per acre, while the ARC maximum level of support remains at $52 per acre and ARC only delivers support for yields up to 26 bushels per acre. Under 2014 price and yield conditions at planting time, the ARC program payment rates provide support "as if" it were an SCO payment rate, with a coverage level between 75 and 80 percent.

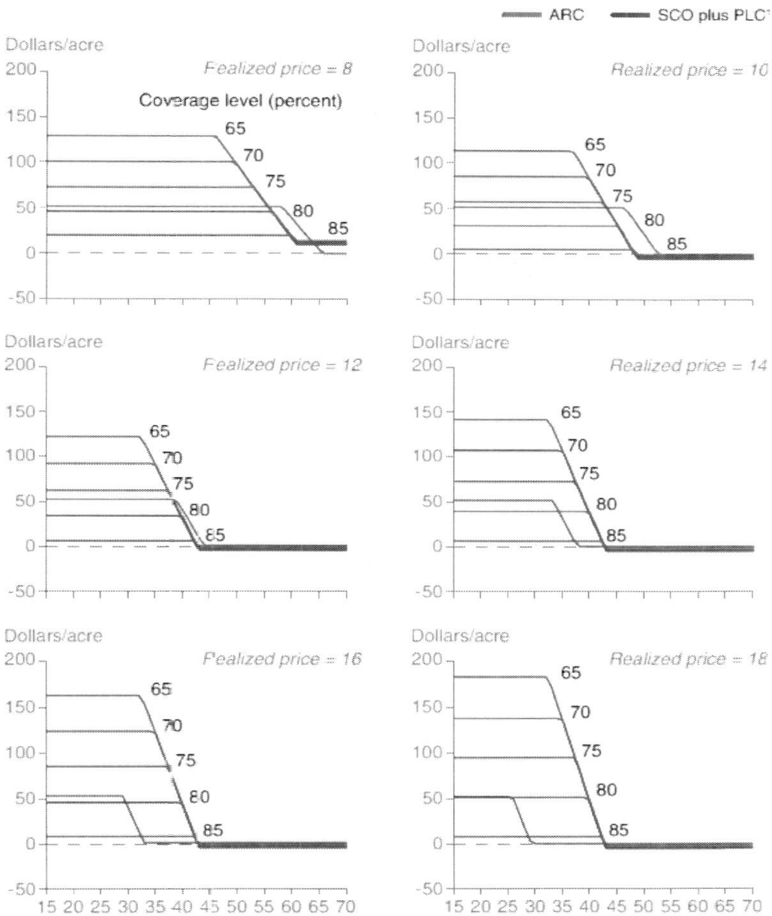

Source: USDA, Economic Research Service.

[1]To calculate the PLC payment, we assume a program payment yield for soybeans equal to 90 percent of the production weighted national yield from 2008 to 2012, which in this case equals 38 bushels per acre. Note that for an individual producer, the program payment yield would be calculated at the farm level. Also note that in the examples shown here, PLC payments only accrue in the top-left chart where the realized price of $8 per bushel falls below the trigger reference price of $8.40 per bushel.

ARC = Agriculture Risk Coverage program; PLC = Price Loss Coverage program; SCO = Supplemental Coverage Option.

Figure 5. Comparing ARC to SCO plus PLC1 payment rates (assuming price and yield expectations for 2014).

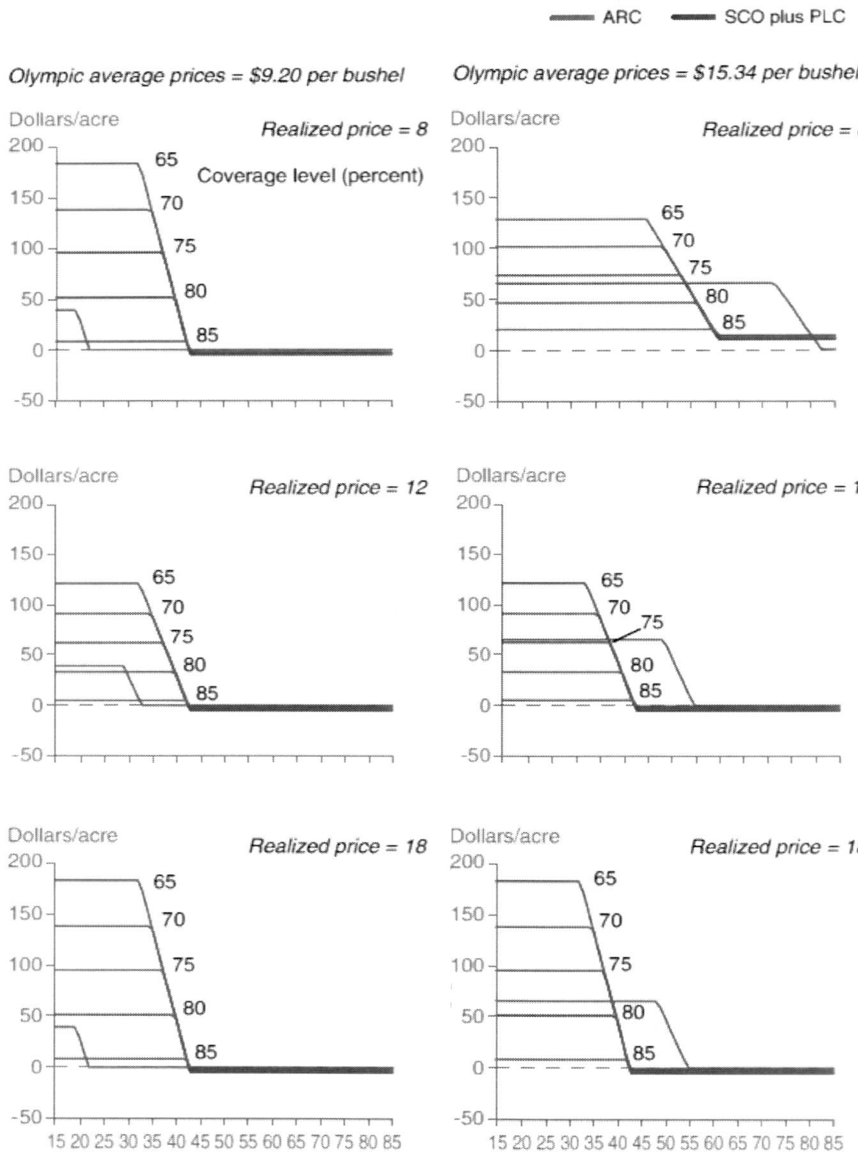

Source: USDA, Economic Research Service.

ARC = Agriculture Risk Coverage program; PLC = Price Loss Coverage program.

Figure 6. Historical prices affect ARC support, but do not alter SCO support.

Note, however, that different initial conditions (if, for example, prices and/or yields went on a multiyear decline) could alter the payment outcome. Figure 6 explores this possibility using two scenarios, the first assuming that historical prices were three-quarters of those reported here ($9.20 vs. $12.27) and the second assuming that historical prices were 1.25 times the current prices ($15.34 vs. $12.27). Changing historical prices has no bearing on the SCO program—note that the SCO charts in figure 6 do not change with the change in historical prices.

Because the benchmark used to determine ARC support levels uses historic prices, ARC does not provide as much support in the lower historic price scenario as in the higher historic price scenario, all else being equal. The difference in ARC program support can be substantial given different historic prices. For example, with historic Olympic-average prices near $9 per bushel and a realized price of $8 per bushel, the ARC program provides support up through realized yields of 48 bushels per acre, while if historic Olympic-average prices were near $15 per bushel, the ARC program provides support up through realized yields of over 80 bushels per acre (fig. 6).

All else being equal, a support program with a slower adjusting revenue guarantee has a greater probability of providing payments when prices are on a falling trend across seasons, but a smaller probability of providing payments when prices are increasing across seasons. Given the recent relatively high commodity prices prior to 2014, it is not a surprise that the USDA, FSA data show that the ARC program appears to be relatively attractive to farmers.

PRODUCER RETURNS UNDER ALTERNATIVE PROGRAM CHOICES

The previous chapter assumed that realized prices and yields were deterministic in nature, that is, they did not include the reality of uncertain future price and yield outcomes. The deterministic approach facilitates the examination of the mechanics of the SCO and ARC programs, exploring the relationship between prices, yields, insurance coverage rates, and payments. This chapter performs a forward-looking analysis by assuming that prices and yields follow stochastic processes, permitting us to examine the distribution of payments and revenue with and without payments that eligible producers are likely to face. Based on an estimated distribution of prices and yields, the expected value (mean) of revenue and payments can be calculated, as can other characteristics of their distributions, such as those related to variability. Since farmers are likely to be risk averse (e.g., Serra et al., 2006), we examine the impacts the program payments would have on measures of revenue variability and expected revenues and also explore how changes to these programs would affect producer revenues. In addition, not all counties have the same correlations between prices and yields, allowing us to explore where the ARC and the SCO programs cause the greatest risk reduction.

The relationship between price and national average yield is generally negative, although the strength of the correlation depends on the crop. The historical correlations for corn, soybeans, and winter wheat are -0.59, -0.50, and -0.32, respectively, based on data from 1975 through 2013. At the national level, low yields often result in high prices and vice versa, creating a "hedging" effect. However, for those counties and States that do not reflect the national yield, this hedging effect may not exist. To see this effect or lack thereof, appendix figures 1-3 are maps that show the correlations between simulated yields of each county's representative farm and the simulated prices. We suspect the greatest benefit of these revenue programs may exist in those counties that do not historically follow the national average simply because these counties do not have the built-in price-yield hedge that the counties that do tend to follow the national average enjoy. For example, counties producing corn outside the Corn Belt would likely see greater risk reduction than counties inside the Corn Belt.

We use a statistical model to generate distributions of simulated prices and yields that are centered around their expected values at planting time in 2014 (see box, "The Empirical Approach"). County-level yield distributions are generated for each of the 1,001 counties for which USDA,

NASS has reported corn data, 889 counties for which NASS has reported soybean data, and 510 counties for which NASS has reported winter wheat data from 1975 through 2013. As a result, not all counties with production are included; 84 percent of 2013 total U.S. corn production, 82 percent of total soybean production, and 56 percent of total winter wheat production are represented in the analysis.

The analysis maintains the historical correlations of yields across counties and between county yields and prices. The yield distribution for a typical farmer in each county is generated by inflating the county-level yield variability based on farm yield information implicit in actual crop insurance-premium rates for each county. The model makes 10,000 draws from each county and farm yield distribution as well as from the price distribution. Payments, crop revenue, and total revenue (crop revenue plus the payments) are calculated for each of the 10,000 price and yield pairs. Producer-paid premiums for SCO and the underlying insurance policy are also calculated from this price-yield data and are subtracted from the payments in order to obtain a "net payment" to the producer. Total revenue then equals the crop revenue plus the net payments. We summarize the results by presenting their expected values and their confidence intervals, the latter being one way to express variability.[5] The results include the costs associated with the various programs, which are the farmer-paid premiums for SCO and the underlying crop insurance policy.[6] With the analysis, we explore the benefits to having: PLC; standalone SCO (i.e., no underlying insurance policy; SCO without an underlying insurance policy is not possible in practice, but we want to allow the direct comparison of the SCO and ARC payments); county-level ARC revenue protection (RP) with a 75-percent coverage rate; RP with a 75-percent coverage rate plus

---

[5] The confidence intervals presented here do not assume any particular distribution of the underlying data (that is, the empirical confidence intervals are nonparametric). While we present the coefficient of variation of total gross revenue in the tables as it is a common statistic, the lower bound of the confidence interval is more informative of downside risk. Using the lower bound value to compare riskiness across programs is similar in concept to the value- at-risk (VAR) approach used in financial risk management.

[6] The net payments are calculated by generating the total premium and using the appropriate premium subsidy rates to calculate the portion the producer pays (the farmer-paid premium). This farmer-paid premium is then subtracted from the total premium to obtain the amount of benefits the producer receives from the program/policy.

SCO and PLC; and RP with a 75-percent coverage rate plus county-level ARC.[7] Although traditional crop insurance is not required to obtain ARC, we examine the effect of the combination of ARC and traditional crop insurance on crop revenue for two reasons: first, as with other major field crops, the bulk of soybean, corn, and wheat acreage in the United States is insured; and second, doing so provides a comparison to returns under SCO and traditional insurance.

Table 1 shows the average of the revenue and payment simulations across representative producers in each county producing either corn, soybeans, or winter wheat. The results show the national weighted average of what a typical U.S. producer would expect to generate in revenues from 1 acre planted to corn, soybeans, or winter wheat under the different support scenarios. For the payments coming from ARC or PLC, the averages reported in the table are generated by weighting the results for each producer by the total commodity base acres in the producer's county. For payments that do not involve ARC or PLC, the weights are total planted commodity acres in the county. If a producer did not enroll in any programs, but simply planted the acres to the crop, the producer would expect to earn a per-acre revenue of $664 for corn, $498 for soybeans, or $261 for winter wheat. For simplicity, following corn only, based on 10,000 simulations, 95 percent of the time the revenue fell within the interval of ($232 to $1,157) suggesting that with a 95-percent probability, the farmer could expect to receive no less than $232 per acre (in bad years) or no more than $1,157 (in a good year) from simply the gross revenue and no program payments—based on price and yield realizations—of the crop.

Of course, this does not rule out the possibility of complete disaster wiping out the crop entirely (for a revenue of $0), or a revenue that

[7] We choose the 75-percent coverage level because it is currently one of the most commonly selected coverage levels. Note that these new programs may induce producers to change their coverage level choices, but for sake of the report, we use the 75-percent level and explore in a couple instances how changes to this level may affect outcomes.

exceeds $1,157 per acre, but the model suggests that such extreme outcomes are likely to happen less than 5 percent of the time.[8]

The Empirical Approach

The approach to the stochastic simulation used in this chapter is described in detail in Cooper and Delbecq (2014a; 2014b), which in turn is an update of Cooper (2010). A county-based simulation model is used, one that was developed to generate stochastic price and yield outcomes that are correlated across space and time. Using historic prices and yields as the basis from which to generate nonparametric kernel distributions of these variables, each run of the model consists of 10,000 price and yield draws. These draws form the distribution of price and yield outcomes used to generate expected price and yield outcomes for every county in the United States growing soybeans for which USDA's National Agricultural Statistics Service (NASS) reports county-level yields from 1975 through 2013 (USDA, NASS).

Since prices and yields are correlated, as are yields across counties, these historic correlations need to be accounted for in the simulated price and yields data. For this analysis, the historic relationships between prices and yields and between county yields are assumed to follow a multivariate Gaussian distribution and are imposed via a copula approach. Under this approach, the historic relationships between the variables are defined by the correlation matrix and a degrees of freedom parameter, which are allowed to vary by ERS Farm Resource Region (Heimlich, 2000).

County-level yields are the lowest aggregation of yield data available from the USDA that has the same time series as the State and national data. However, farm-level yield tends to exhibit higher variability than county-level yield. As per Coble and Dismukes (2008), the additional variability at the farm level for each county is inferred from Federal crop insurance premiums, and the variability of the simulated county-level data is increased by this amount. The result is a yield distribution for a representative farmer in each county. The generated 2014 farm yields and harvest time and season average prices are then used as inputs into the various programs offered (ARC, SCO. etc.) to examine potential payment and revenue outcomes for representative producers within these counties. For the national-level average payments, depending on the support program, the revenue and payment outcomes are weighted by the number of base acres or planted acres in the county.

[8] Note that the 95-percent percent upper bounds with RP and RP plus SCO are lower than the bounds without program support (the "unenrolled" column) or with ARC only because the RP and SCO calculations include the farmer-paid premiums for these products.

If a producer enrolled only in the PLC program, the average total revenue would increase to $675 per corn acre with a 95-percent probability of the total revenue for a given year falling within the range of $242 and $1,164. PLC limits losses due to drops in prices, reflected in an increase in the lower bound of the 95-percent confidence interval of $3 per acre (from $239 to $242 per acre).[9]

While a producer cannot enroll only in SCO (he or she must have an underlying insurance policy to be eligible to enroll in the SCO program), we include the possibility of only enrolling in SCO to enable a direct comparison with the ARC program as well as exploring the benefits accruing directly from the SCO program itself. Simulation results suggest that under the 2014 price scenarios used here, a producer with SCO coverage would expect an average total revenue of $677, which includes a cost (of roughly $13) for enrolling in the program. With a 95-percent probability, the producer could expect a revenue between $281 and $1,154 for each corn acre planted. Note that the upper end of the revenues is slightly lower due to the producer's cost of enrolling in the program. Raising the lower end of revenue, however, is the likely reason a producer interested in risk management would enroll in the program—to limit downside risk and the associated losses. As a result of enrolling in SCO only (again, a choice not actually available), the lower end of the 95-percent confidence interval now lies at $281 per acre—$49 higher than with not enrolling in any programs at all.

If a producer enrolled only in ARC, the program provides a slightly higher expected per-acre net payment (ARC has no enrollment fees; the ARC payment is based on base acres, not planted acres) of $699 per acre to the producer, but appears to limit downside risk in this scenario (under current historic prices and yields) slightly more than the SCO (plus PLC) program, with an associated lower end of the 95-percent confidence interval of $296, an increase of $15 per acre over the lower bound found with the SCO-only election.

[9] Recall that the PLC and ARC program payments are weighted by base acres, not planted acres, resulting in the higher base levels of gross revenue.

Table 1. Per-acre payments and revenues for producers in all U.S. corn-, soybean-, and winter wheat-producing counties (2014 expected prices and yields)

Farmers' program enrollment	Average (net) payment ($/acre)	Average total revenue ($/acre)	95-percent confidence interval of revenue ($/acre)	Coefficient of variation of total revenue/acre[1]
Corn				
Not enrolled-planted[2]	--	664	[232, 1157]	0.36
Not enrolled-base[2]	--	671	[239, 1164]	0.36
PLC only[2]	4	675	[242, 1164]	0.35
SCO only[3]	13	677	[281, 1154]	0.34
ARC only[2]	28	699	[296, 1167]	0.32
RP (75 percent)	23	687	[494, 1138]	0.27
RP + SCO + PLC[2]	39	710	[501, 1143][4]	0.25
RP + ARC[2]	50	721	[517, 1148][4]	0.24
Soybeans				
Not enrolled-planted[2]	--	498	[151,910]	0.40
Not enrolled-base[2]	--	507	9 1 1]	0.38
PLC only[2]	0.17	507	9 1 1]	0.38
SCO only[3]	9	507	[177, 909]	0.38
ARC only[2]	13	520	[195, 912]	0.36
RP (75 percent)	19	517	[359, 894]	0.30
RP + SCO + PLC[2]	26	533	[362, 895][4]	0.28
RP + ARC[2]	31	538[a]	[367, 897][4]	0.27
Winter wheat				
Not enrolled-planted 2	--	261	5 5 6]	0.55
Not enrolled-base[2]	--	276	[31, 589]	0.54
PLC only[2]	2	278	5 8 9]	0.54
SCO only[3]	7	268	[50, 555]	0.52
ARC only[2]	4	280	[43, 590]	0.52
RP (75 percent)	16	278	[180, 543]	0.40

Table 1. (Continued)

Farmers' program enrollment	Average (net) payment ($/acre)	Average total revenue ($/acre)	95-percent confidence interval of revenue ($/acre)	Coefficient of variation of total revenue/acre1
RP + SCO + PLC2	26	302	[187, 574]4	0.36
RP + ARC2	21	297	[190, 575]4	0.38

Source: USDA, Economic Research Service.

ARC = Agriculture Risk Coverage program; PLC = Price Loss Coverage program; RP = Revenue Protection; SCO = Supplemental Coverage Option.

[1] "Coefficient of variation of revenue" is the standard deviation of revenue divided by average revenue, and is a standardized measure to allow comparability of variability across programs.

[2] With ARC and PLC, the results across counties are weighted by base acres ("base" in the first column), while SCO and RP are weighted by planted acres ("planted). Also, note that the confidence interval is nonparametric (i.e., does not assume any particular distribution), and as such, and may not be symmetric around the average.

[3] "SCO only" presumes a 75-percent coverage rate on the RP policy.

[4] While these are likely close to the "true" interval, caveats must be made due to their construction. They are the result of averaging across counties while weighting by base acres, which can cause funny outcomes. Adding ARC (or SCO plus PLC) to RP will not cause the lower end of the distribution to change for any individual county (see fig. 11 for details). It will shift many producers to higher levels of revenues, indicated by the increase in the average total revenue, and will lower the CV of revenue, and can reduce downside risk as measured by other measures, such as semi-variance (used in the finance industry), but will not shift the lower end of the distribution because these are area-based policies/programs and as a result, will not provide benefits for everyone suffering losses. There will be enough who do not correlate perfectly with the county outcomes such that the lower bound of the 95-percent confidence interval will, in almost all cases, remain unchanged. With respect to the upper bound, it is also not clear how RP+SCO+PLC can have a higher upper bound than RP alone given SCO also has a premium associated with it, which should lower that upper bound.

If the producer decided to enroll in the Federal Crop Insurance (FCI) program with a revenue policy and 75-percent coverage, the RP policy would net the farmer an average of $23 per acre. This would increase the average total revenue to $687 for each corn acre planted. The producer has a positive net average benefit from the RP policy, as the Government is paying a portion of the premium.[10] With a 95-percent probability, a

[10] For the sake of internal consistency across gross revenue, ARC, SCO, and RP calculations, the actuarially fair (e.g., Arrow, 1963) SCO and RP insurance premiums are set to the expected value of the indemnities across the 10,000 price-yield outcomes. The producer-paid premium excludes the Government-paid portion (i.e., the premium subsidy).

producer could expect a revenue falling between $494 and $1,138. The lower end of the 95-percent confidence interval now lies more than $260 per acre higher than the lower bound of the associated interval for the no-insurance outcome and roughly $200 above the lower bound of either the SCO- or ARC-only outcomes. This is why SCO and ARC are colloquially referred to as “shallow loss” programs. They are not designed to cover the deeper losses that the RP program does cover.

When SCO (plus PLC) and ARC are separately combined with an underlying RP policy with 75-percent coverage, the resulting expected level of revenue and range of likely revenue outcomes are close for the 2014 planting-time price scenario, suggesting that the two county-based programs work similarly under the 2014 scenario of historic and expected yields and prices. Overall, when looking across all counties, under the 2014 scenario, ARC appears to generate higher mean levels of revenue, and the lower end of the 95-percent confidence intervals appear greater as well for all three crops.

Underlying Revenue Protection Coverage Levels

The results presented above for the RP, the RP plus SCO plus PLC, and the RP plus ARC options all assumed that the producer would select an underlying RP policy with 75-percent coverage (the most common choice currently selected by producers). However, the introduction of SCO (plus PLC) and ARC may cause producers to change their coverage election of the underlying insurance policy. In other words, rather than these shallow-loss programs acting as complements to the FCI program, they may act as substitutes. To explore this possibility, we examine the corn revenue distributions for corn for RP coverage selections of 65 and 55 percent, respectively, to compare them to the benchmark election of 75 percent that represents the coverage level most commonly elected before the advent of these programs (table 2).

The underlying coverage level of RP affects the payouts of the various combinations of programs selected by the producers. While the average revenue per acre does increase over the various coverage levels selected, the lower bound of the confidence interval changes more substantially.

Table 2. Changing the underlying coverage level of RP plan can alter revenue distributions[1]

Farmers' program enrollment	Average (net) payment ($/acre)	Average total revenue ($/acre)[2]	95-percent confidence interval of revenue ($/acre)[3]	Coefficient of variation of total revenue/acre[4]
Not enrolled-planted	--	664	[232, 1157]	0.36
Not enrolled-base	--	671	[239, 1164]	0.36
55-percent RP coverage				
RP (55 percent)	9	674	[372, 1152]	0.32
RP + SCO + PLC	34	705	[415, 1156]	0.28
RP + ARC	37	708	[427, 1162]	0.28
65-percent RP coverage				
RP (65 percent)	15	679	[434, 1146]	0.30
RP + SCO + PLC	37	708	[452, 1151]	0.26
RP + ARC	42	713	[474, 1157]	0.26
75-percent RP coverage				
RP (75 percent)	23	687	[494, 1138]	0.27
RP + SCO + PLC	39	710	[501, 1143]	0.25
RP + ARC	50	721	[517, 1148]	0.24

Source: USDA, Economic Research Service.

ARC = Agriculture Risk Coverage program; PLC = Price Loss Coverage program; RP = Revenue Protection; SCO = Supplemental Coverage Option.

[1] Results across a representative producer in each U.S. corn-producing county, given 2014 expected prices and yields.

[2] With ARC and PLC, the results across counties are weighted by base acres, while SCO and RP are weighted by planted acres.

[3] Note that the confidence interval is nonparametric (i.e., does not assume any particular distribution), and as such, may not be symmetric around the average.

[4] "Coefficient of variation of revenue" is the standard deviation of revenue divided by average revenue, and is a standardized measure to allow comparability of variability across programs.

For example, with a 75-percent underlying RP policy, having SCO and PLC as well generates a revenue distribution with a mean of $710 per acre; with a 55-percent underlying RP policy and the same combination of programs, the mean drops to $705 per acre, a relatively minor drop. However, the lower bound of the confidence interval decreases from just over $500 to $415 per acre, signifying that if a loss occurs, it has a greater probability of being a larger loss with the lower underlying coverage level. Recall that even though SCO will cover losses down to the level of the underlying policy, and will do so at the same rate regardless of the coverage level selected for the underlying policy, the SCO is an area-based policy, so the degree of protection afforded to the producer relies heavily on how correlated an individual producer's returns are to those of the county where the farm is located. Therefore, these results may differ by county. In addition, the upper bound of the 95-percent confidence interval changes when moving to the lower RP coverage level, increasing from $1,143 to $1,156 per acre, reflecting the additional out-of-pocket costs (of the premium) the producer faces when enrolling in the RP and SCO policies. Exploring corn, for the lower coverage levels, ARC versus SCO and PLC provides roughly the same decrease in risk as measured by the coefficient of variation (CV).[11] At 75-percent coverage, however, ARC appears to provide slightly more protection against risk. At an aggregate level, examining across all counties, the 75-percent RP coverage coupled with ARC provides the highest level of risk protection for a corn producer. This pattern holds for soybeans as well, while for winter wheat, the 75-percent RP coverage coupled with SCO provides the highest level of risk protection. This result helps explain why wheat producers were more likely to elect PLC than ARC, according to the FSA election results. Note, however, that these results are likely to be sensitive to the base year chosen for the scenario.

---

[11] The coefficient of variation (CV) is calculated by dividing the standard deviation by the mean of the variable, relating a measure of variability to the population's average level of the variable. The larger the estimated CV, the higher the variance relative to the mean, which denotes a higher level of riskiness.

Different Levels of Revenue Risk Produce Varying Results

To explore the idea of whether results vary for different levels of revenue risk being faced by producers (and if so, how), three specific counties are examined where producers face various levels of risk producing soybeans—Linn County, IA; Stearns County, MN; and Prentiss County, MS. First, we calculated the CV for every county growing soybeans in the United States. All soybean counties fell between a low of 0.25 and a high of 0.82, where the lower the CV, the lower revenue risk the producer faces. Linn County, IA, is typically a low-risk production area, with a CV of 0.28. Stearns County, MN, represents a medium-risk county with a CV of 0.55, while Prentiss County, MS, is a high-risk county for soybean production with a CV of 0.76 (see the figure 7 map for where these counties are located). Figure 8 graphically represents the expected (average) revenue—denoted by the dot—and the range of revenue a soybean producer could expect to fall within 95 percent of the time—denoted by the whiskers—associated with the various programs the producers could enroll in (the data for figure 8 resides in appendix table 8 as well). We also include "all counties," the acreage-weighted average results across all U.S. counties for which USDA, NASS has reported soybean production over 1975-2013, as a comparison to obtain a sense of the overall average effect of enrolling in the various programs.

Overall, across all counties included in the simulation model at planting time in 2014, a soybean producer could expect to receive roughly $500 per acre. If not enrolled in any Federal programs, the expected revenue would lie somewhere between roughly $150 and $910 per acre 95 percent of the time. Enrolling in the support programs reduces the lower end of the distribution—in effect, shortening the lower whisker in the figure. However, overall averages often mask underlying variation. A typical soybean producer in Stearns County, MN, expects a lower revenue, on average, than a typical soybean producer in Linn County, IA (by roughly $100 per acre). Furthermore, a producer in Stearns County, MN, is more likely to lose a substantial amount, if not the producer's entire expected revenue, compared with a typical producer in Linn County, IA. This suggests that Stearns County is a riskier place to grow soybeans than in Linn County, and the relatively larger benefit from enrolling in the various safety net programs is evident by the degree to which the lower "whiskers" are shortened—implying a greater reduction in low-revenue outcomes. Following our expectations given the calculated CVs, results suggest that Prentiss County, MS, is even riskier yet. Producers in both Stearns and Prentiss counties on average expect lower levels of revenue and stand to lose more than a producer in Linn County, IA, and therefore have greater changes in the lower bound of revenue with enrollment, particularly regarding RP.

Note also that in Linn County, IA, producers would face less downside risk if they adopt standalone SCO or ARC.[12] As the risk level increases, however, the benefits from these programs in terms of alleviating downside risk decreases. Producers in a very high-risk county like Prentiss County, MS, do not receive many downside risk reduction benefits from enrolling in either standalone ARC or SCO (if they could indeed enroll in SCO alone). For the riskiest county, the downside risk protection is negligible—the lower bound remains 0 (and even goes negative for standalone SCO—since producers have to pay a premium) despite enrolling in these programs. These differences among counties in the standalone benefits of SCO and ARC are due to small revenue losses being more typical in lower risk Linn County, IA, than in the two more risky counties.

When combining SCO or ARC with RP, the additional downside risk protection—defined as the increase in the lower bound of the 95-percent

---

[12] Again, it is not possible to obtain standalone SCO. The point, however, is to compare the SCO with the ARC program in the absence of traditional crop insurance to examine which program provides more downside risk protection when dealing with a riskier environment.

confidence interval on revenue—disappears, as shown in the figures and in appendix table 8. However, this does not mean that adding SCO or ARC in conjunction with RP does not reduce downside risk. For example, if calculating the semi-variance, a measure of downside risk that attempts to capture the variability of "poor" or undesirable returns (e.g., the variation of returns that are below the average return) used in the finance literature, adding SCO or ARC to RP does decrease such a measure of downside risk.

WHAT THE NEW DISTRIBUTIONS LOOK LIKE

Figure 9 allows us to compare the various distributions emerging from the different programs farmers can enroll in for a low-risk county (Linn County, IA) and provides a better idea of how the various programs deal with risk in terms of the revenues and the probability of a producer realizing a revenue in a particular part of the distribution. The X axis shows the range of revenues while the Y axis denotes the density associated with a particular revenue.[13] The chart was made using the revenue data from each of the 10,000 simulations run for a typical producer in this particular county.

The gold curve shows the crop revenue distribution when the producer does not enroll in any programs. The distributions for the crop revenue plus the crop insurance products (RP; RP plus ARC; and RP plus SCO plus PLC) look fairly similar. RP limits downside risk by effectively cutting off the left tail of the revenue distribution (denoted by the dark orange distribution). With a 75-percent coverage choice, the RP policy guarantees at least 75 percent of expected farm revenues (less the farmer-paid RP premium). As a result, a large spike occurs at the 75-percent level of

[13] The density is related to frequency of occurrence, with the higher the density, the higher the frequency of the associated revenue occurring. The area under the density function sums to 1, that is, 100 percent of all occurrences are covered under the graph.

expected revenues, which is the result of all the producers who had losses that brought their revenue below the 75-percent level, which their RP policy then made up for—up to the 75-percent level. To the right, the spike quickly drops and then comes close to following the revenue distribution.[14]

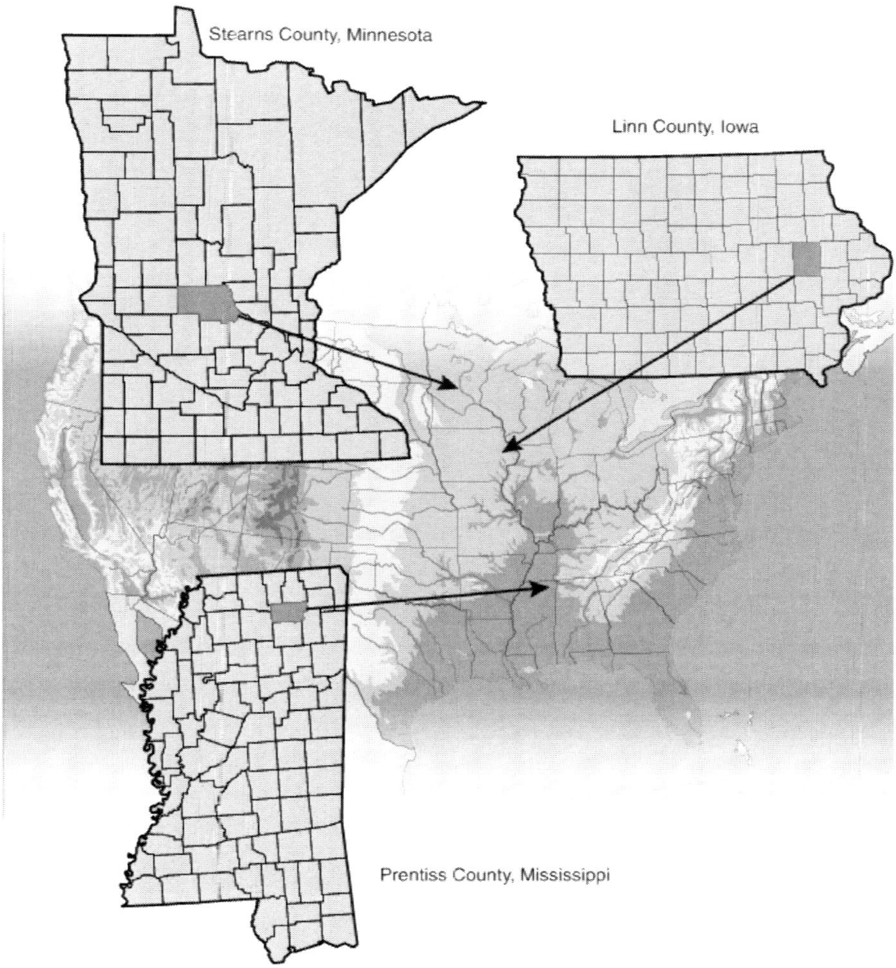

Source: USDA, Economic Research Service.

Figure 7. Location of Linn County, IA; Stearns County, MN; and Prentiss County, MS.

[14] The reason that the spike is not a thin vertical line is due the limitations of the sampling of the distribution being finite.

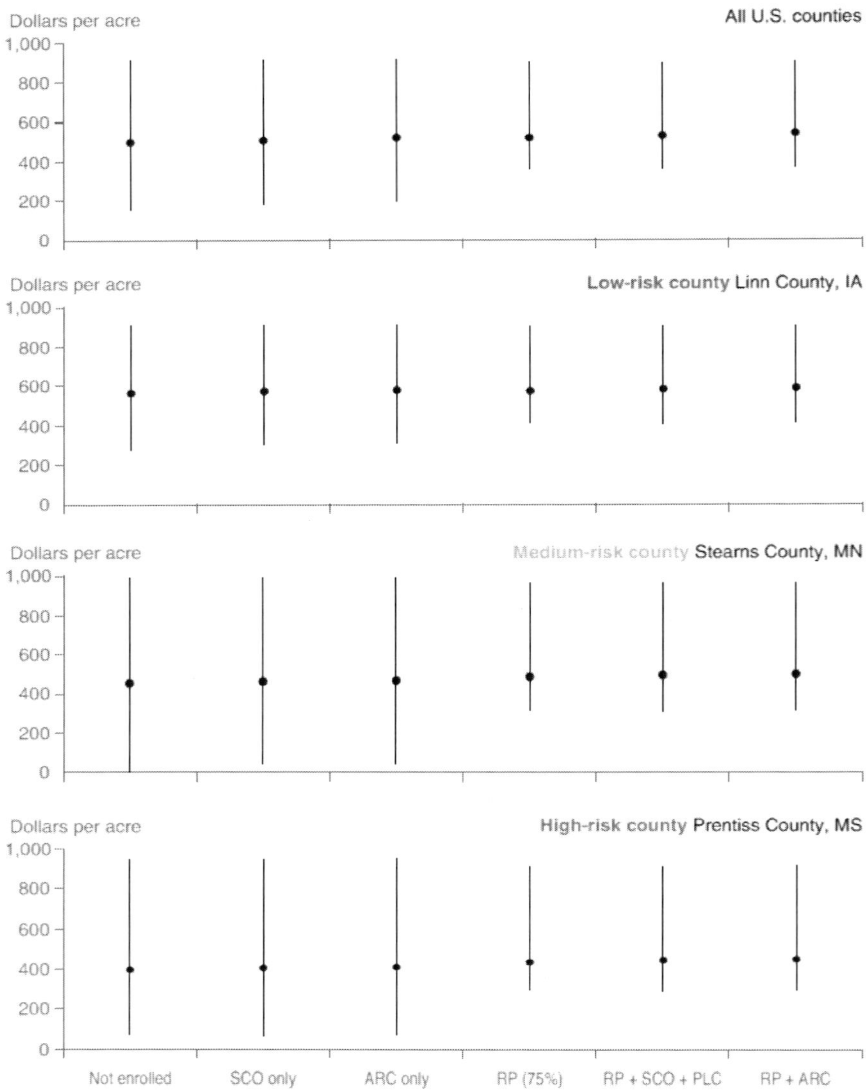

Source: USDA, Economic Research Service.

ARC = Agriculture Risk Coverage program; PLC = Price Loss Coverage program; RP = Revenue Protection; SCO = Supplemental Coverage Option.

Figure 8. Expected soybean revenue outcomes, all U.S. counties and county examples by level of revenue risk.

The blue line shows the revenue distribution with RP, SCO, and PLC in place. There is a similar, but much smaller spike at the left side of this distribution, due to the fact that in any scenario with RP, revenue cannot be lower than the revenue protection guarantee (less the farmer-paid premium) under RP. The reason that the leftmost spike is lower than in the RP-only case is that a second spike occurs close to the 86-percent level of expected revenues, since this is where SCO begins to make its payments for the smaller losses between the RP coverage rate and 86 percent of expected county revenues (the area under the line sums to 1, meaning an increase in height in one part of the distribution must be offset by a lowering somewhere else in the distribution). Note also that the left "tail" of the blue distribution lies slightly to the left of the dark orange RP-only distribution. This is due to the fact that producers incur a premium in order to purchase SCO and, since SCO is an area-based product, producers could pay for SCO, incur losses, and if those losses do not coincide with losses at the county level, can leave them with no indemnity payments, despite having incurred losses and paid a premium. The "lost premium" is what moves this distribution to the left of the RP-only distribution. Given expected prices, the PLC program is not expected to alter the distribution much.

The green line denotes the revenue distribution with RP plus ARC. This has a similar shape to that of the RP plus SCO and PLC curve, although it has a lower first spike and a higher second spike. Because there is no premium associated with the ARC program, the green curve lies to the right of the blue curve and above it as revenues increase. The second spike is higher, suggesting that for soybean producers in Linn County, IA, a low-risk county, the ARC program may provide higher levels of benefits than the SCO program, with a higher likelihood of a producer obtaining a larger revenue (evidenced by the higher second spike to the right and a lower spike to the left at the RP cutoff).

Finally, as expected, all distributions converge as we move to the right along the revenue axis. As producers realize higher levels of revenues, the programs do not tend to pay out indemnities—with the potential exception of the RP policy and associated SCO, which provides some upside price protection given that the base price is the higher of the planting time or harvest price. However, in any scenario, higher yields leading to higher revenues tend not to trigger any payments. As a result, the distributions get too close together to be able to discern any meaningful differences between the four distributions.

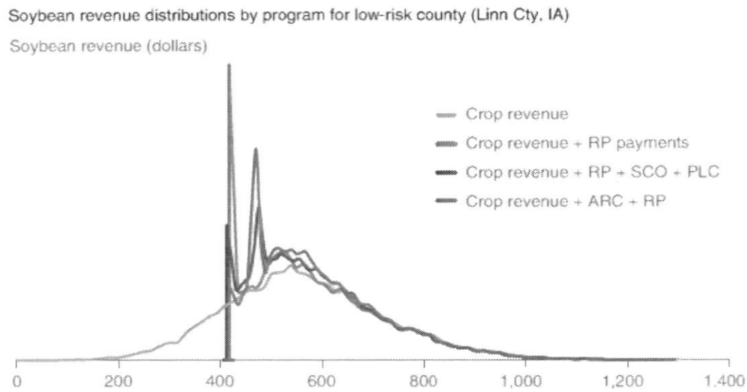

Source: USDA, Economic Research Service.

ARC = Agriculture Risk Coverage program; PLC = Price Loss Coverage program; RP = Revenue Protection; SCO = Supplemental Coverage Option.

Figure 9. Example of impacts of RP and SCO on the distribution of soybean revenue for a low-revenue-risk county.

NEW PROGRAMS' RISK COVERAGE RELATIVE TO TRADITIONAL CROP INSURANCE

Understanding how the SCO and ARC programs reduce producers' risk above and beyond that already covered by traditional crop insurance remains difficult. Relative to the benefits of traditional crop insurance, how large are the added benefits these programs provide? And do these relative benefits differ for producers facing different levels of risk? In other words, are SCO and ARC more valuable relative to the underlying crop insurance policy for some producers than others?

To explore this concept, we generate measures of the total costs of the SCO and ARC programs relative to the total costs of the underlying traditional RP crop insurance policy. The total costs reflect the amount the Government has to pay to provide the support to the producers as well as the producer's payment; the total captures the level of risk being covered by each type of support. For SCO and RP, we define total costs as the total (farmer- plus Government-paid) premiums for these programs. The total premiums embody the total cost on average of providing these supports. The total cost of ARC is the expected payment, ignoring administrative costs. We then create a ratio of the total costs of either SCO or ARC to the RP policy. If close to 0, this would suggest that the SCO (ARC) program provides little, if any, additional value to the producer in terms of revenue risk covered above and beyond the benefits of the underlying RP crop insurance policy. If close to 1, SCO (ARC) provides value very close to that of the underlying policy. If above 1, the SCO (ARC) program provides value that exceeds that of the underlying policy. We then graph these ratios of relative benefits according to the riskiness of the producer (measured by the CV of revenues).

Figure 10 shows these ratios in graphical form. A clear relationship emerges showing that the ratio tends to be higher when the farmer's revenue risk is lower. As we suspected from comparing the three counties earlier in the report, this graph demonstrates that the shallow-loss programs are relatively more important for low-risk producers than they are for high-risk producers. This makes sense since producers in low-risk areas of the country have a low probability of incurring large losses. The primary losses they face tend to be smaller in nature and the shallow-loss programs help with exactly these types of losses. For these lower risk producers, the shallow-loss programs are almost as important, if not more important, in terms of mitigating the risk they face when compared to the traditional, underlying RP policy. For example, Linn County, IA, a low-risk part of the country, has an SCO-to-RP ratio that exceeds 1, suggesting that producers in Linn County would, on average, obtain benefits from the SCO program greater than those of the underlying RP program (fig. 10a). In higher risk counties, like Prentiss County, MS, the relative benefit is small. Producers in these riskier parts of the country are more worried about the large losses they have to face and the benefits of the shallow-loss programs—SCO and ARC—are relatively small when compared to the benefits received from the traditional underlying RP policy.

Under 2014 conditions, the underlying riskiness of the county (measured by the county's CV of gross revenue) explains more of the variation in the ARC/RP ratio than in the SCO/RP ratio.[15] However, a 1-percent increase in the CV causes a smaller marginal percentage drop in the ratio for SCO than for ARC, suggesting that SCO maintains a higher level of benefits than does ARC when production becomes more risky. Focusing solely on this metric, SCO generally appears more attractive than ARC to typical producers. This could, in part, be due to the fact that producers receive some upside-price protection with SCO (given the assumption of an underlying RP policy) that is absent with ARC.

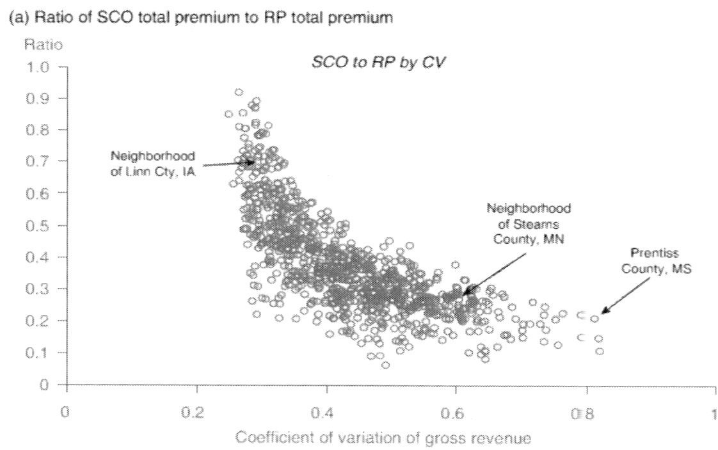

Figure 10. (Continued).

[15] Using simple double-log univariate regressions of the data in figure 8, 59 percent of the variation in the SCO/RP ratio is explained by the CV of gross revenue and 77 percent of the variation in the ARC/RP ratio is explained by the coefficient of variation of gross revenue. Note that the denominator (RP) is increasing in the coverage-rate choice, SCO is decreasing in the RP coverage rate, and ARC payments are invariant to the coverage choice. Hence, the relationship of the SCO/RP ratio to the CV of revenue is likely to have some sensitivity to the RP coverage rate choice, while the ARC/RP ratio is strictly decreasing in the coverage choice.

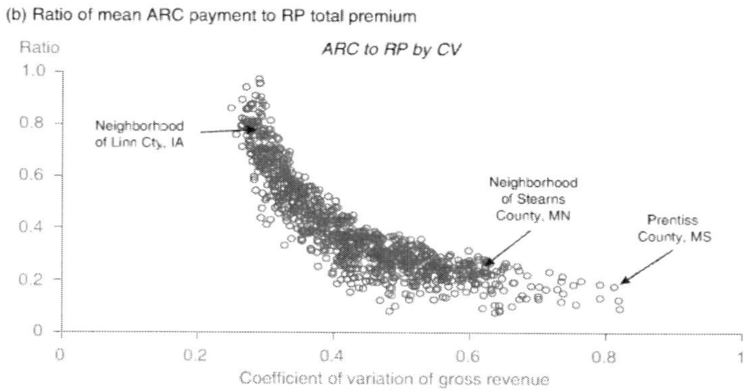

Source: USDA, Economic Research Service.

ARC = Agriculture Risk Coverage program; RP = Revenue Protection; SCO = Supplemental Coverage Option. Coefficient of variation of gross revenue is a measure of risk evaluated by dividing the standard deviation (spread) of gross revenue by its mean (average) value.

Note: Fig. 10 shows a typical producer for each U.S. county for which USDA's National Agricultural Statistics Service reported soybean production from 1975 through 2013.

Figure 10. Ratio of shallow loss to insurance support versus the coefficient of variation of gross revenue for all counties (based on 2014 expected prices and yields and RP with 75-percent coverage rate).

However, since the two shallow-loss programs operate differently, producers may have preferences over the various properties of the two offered programs that will help dictate their final enrollment choices. And, indeed, election results made available from USDA's Farm Service Agency have shown that producers appear to have overwhelmingly elected ARC over PLC (and hence, SCO).

THE ROLE OF PRICE CHANGES IN PROGRAM CHOICE

Program benefits can differ based on how prices change over time (table 3). For example, if the average futures price at planting time—our expected price (E[P])—was high relative to the actual 2014 planting time price, then SCO would have a higher probability of providing a greater level of support to producers than ARC. The SCO revenue-guarantee price equals E[P], while the ARC benchmark price is an Olympic average of season-average prices in the prior 5-year function (with the PLC reference price as a floor). This design characteristic means that the SCO revenue guarantee will change more from year to year than the ARC revenue guarantee (Effland et al., 2014). It also means that changing E[P] for a particular crop year does not change the ARC revenue guarantee in a particular crop year, increasing the sensitivity of ARC payments to within-season price changes relative to SCO. In particular, we assume that the distribution of harvest-time prices will be centered around the planting-time prices. While an actual draw from such a distribution may not be close to the planting-time price, when generating 10,000 draws, an average over these draws will place the price in a reasonably close neighborhood to the planting time price. For example, a 5-percent increase in the expected price at planting time means that all simulated prices increase by 5 percent, and a 5-percent decrease would lead to all simulated prices decreasing by 5 percent. So when expected prices are relatively high, ARC will not provide high payouts, while SCO has upside price protection that will generate higher support for producers. Hence, while ARC raises the lower bound of revenue more than does standalone SCO in the low base-price scenario, the downside risk protection of ARC in the high base-price scenario is essentially nonexistent. Risk reduction is notably better with SCO in the high base-price scenario due to the base price being the higher of the base price or the harvest price, based on our assumption of the underlying traditional policy being RP.Table 3 uses soybeans to show the greater sensitivity of average ARC payments to E[P] that are 50 percent and 150 percent of the 2014 E[P] relative to SCO, in addition to their impacts on revenue (see appendix table 9 for additional results for a low-, medium-, and high-risk county). For example, in the low-price scenario, ARC provides an average of $41 in benefits, while in the high-price scenario, ARC only provides $1 in benefits. With the SCO policy, a low expected price immediately adjusts the guarantee, and SCO ends up only providing $4 in benefits. However, SCO provides roughly three times that amount in a high expected-price scenario due to incorporating an upside price-risk component, assuming RP as the underlying insurance policy.

To better visualize the downside risk protection provided by ARC and SCO, figure 11 graphically compares the change in the lower bound of the 95-percent confidence interval of revenue for different expected price scenarios. For example, the lower bound of the 95-percent confidence interval of gross revenue (no program payments) for the low expected price is $83 per acre under the ARC scenario, which increases to $128 per acre when including program benefits—an increase of $45 per acre. For the high expected price, it is the difference between $249 per acre (the low end of the gross revenue per acre) and $253 per acre (the low end of the gross revenue plus payment per acre), which amounts to $4 per acre. Figure 11 graphically captures these differences in the lower bound of the 95-percent confidence interval between the per-acre gross revenue and the per-acre gross revenue plus program payments (the data can be seen in appendix table 10). This effectively shows how the program affects the lower bound of revenue and how expected prices affect how the programs work. By tracking this change in the lower bound of the confidence interval, we can obtain a good sense of how the programs affect the downside risk the producers face.

If expected prices are low—we use half of the current expected prices—the chart shows that if a producer enrolls in the ARC program, the producer will increase his or her downside risk protection and the (95-percent) lower bound of revenues will increase by $45 per acre. If the producer selects SCO instead, his or her lower bound will increase by only $13 per acre. Under this scenario of low expected prices (and current historical prices and yields), ARC appears to provide greater downside-risk protection. However, as the expected price increases, the SCO program begins to provide more downside protection while ARC provides less. When expected prices are 150 percent of current expected prices, SCO increases the lower bound by almost $40 per acre, while ARC increases it by less than $5 per acre. Clearly, the two programs affect downside risk in different ways, and the level of expected prices can play a large role in determining which program provides the highest level of protection.

Table 3. Sensitivity of mean ARC and net SCO payments—with and without gross revenue—to expected prices (E[P]) being 50 percent and 150 percent of the 2014 expected price (for soybeans, average across all U.S. counties)[1]

ARC			**SCO** *(assuming underlying 75-percent coverage revenue policy in place)*		
Lower E[P] *$5.68/bu*	E[P] *$11.36/bu*	Higher E[P] *$17.04/bu*	Lower E[P] *$5.68/bu*	E[P] *$11.36/bu*	Higher E[P] *$17.04/bu*
Mean payments per acre					
41 (8 ; 47)	14 (0 ; 47)	1 (0 ; 10)	4 (-2 ; 29)	9 (-5 ; 58)	13 (-7 ; 86)
Gross revenue per acre					
253 (83 ; 455)	507 (166 ; 911)	760 (249 ; 1,366)	249 (76 ; 455)	498 (151 ; 910)	748 (227 ; 1,365)
Gross revenue plus payment per acre					
294 (128 ; 493)	520 (195 ; 912)	761 (253 ; 1,366)	253 (89 ; 454)	507 (177 ; 909)	760 (266 ; 1,363)

Source: USDA, Economic Research Service.

ARC = Agriculture Risk Coverage program; SCO = Supplemental Coverage Option; bu = bushel.

[1]"All U.S. counties" represents the analysis of a representative farmer in all counties for which USDA's National Agricultural Statistics Service has reported soybean yield data each year from 1975 through 2013, which is 889 counties. In the data summaries, the results for each farmer are weighted by the number of planted or base acres in the county. The different weights being used is the reason why the "gross revenue per acre" differs for ARC and SCO.

While the commodity-futures prices at planting time have a tendency to alter the level of payments, the historic prices are also important for determining program benefits for the ARC program. These prices help determine the benchmark and guarantee revenues for the ARC program. Under the simulation scenarios explored so far, the Olympic average of historic prices (over 2009-13) was higher than prices at planting time in 2014, which means that there is both an increased likelihood of ARC providing support and that support is likely to be higher due to a relatively high benchmark revenue. (Of course, the reverse would hold true if historic prices as manifested in the ARC benchmark price were lower than the 2014 planting time price.)

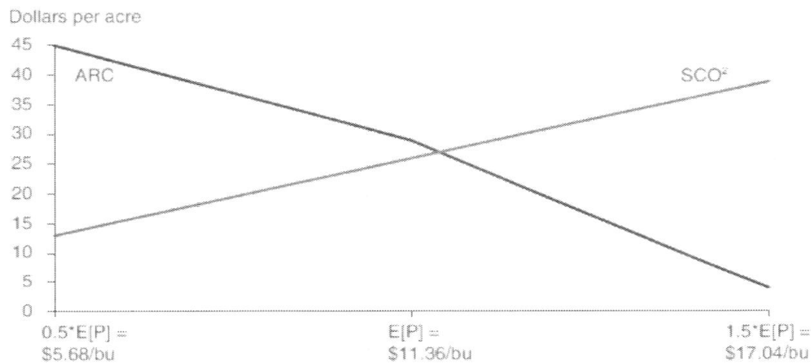

Source: USDA, Economic Research Service.

Note: The $/acre values represent the difference between the lower bound of a 95-percent confidence interval for revenue with the program (ARC or SCO) payment and the lower bound of a 95-percent confidence interval without the payment.

ARC = Agriculture Risk Coverage program; RP = Revenue Protection; SCO = Supplemental Coverage Option.

[1]Enrolling in ARC limits losses more than enrolling in SCO when expected prices are low. If expected prices are high, the reverse holds (E[P]) = 50 percent, 100 percent, and 150 percent of the 2014 expected price.

[2]"SCO" is the SCO payment less what the farmer paid in SCO indemnity and, although examined as a "stand alone" policy here to compare directly to ARC, acts "as if" a 75-percent coverage Revenue Protection policy is in place.

Figure 11. Current environment of prices and yields for soybeans, average across all U.S. counties.[1]

Table 4 explores how different historic prices would affect the program payments made through the ARC program (note that the SCO indemnity payments are not affected at all by changes in historic prices—although SCO premiums can be affected to the extent these historic price changes affect loss ratings). We expect that the impacts on ARC payments and the ARC benchmark price will be the opposite of changing the current season price and, in fact, that is what table 4 shows.

Table 4. Sensitivity of soybean ARC payments to the ARC benchmark price (dollars per bushel)[1]

	ARC payments per acre		
	Low benchmark price	High benchmark price	2014 ARC benchmarkprice
	75 percent of 2014 benchmark price	125 percent of 2014 benchmark price	
All U.S. counties	1.41 (0 ; 18)	41 (0 ; 58)	14 (0 ; 47)
Low-risk county (Linn Cty, IA)	0.51 (0 ; 2)	46 (0 ; 65)	13 (0 ; 52)
Medium-risk county (Stearns Cty, MN)	1.10 (0 ; 26)	37 (0 ; 53)	14 (0 ; 42)
High-risk county (Prentiss Cty, MS)	3.42 (0 ; 23)	31 (0 ; 39)	15 (0 ; 31)

Source: USDA, Economic Research Service.

Note: Numbers in parentheses are lower and upper bounds of a 95-percent confidence interval.

ARC = Agriculture Risk Coverage program.

[1]"All U.S. counties" represents the analysis of a representative farmer in all counties for which USDA's National Agricultural Statistics Service has reported soybean yield data each year from 1975 through 2013, which is 889 counties. In the data summaries for the ARC analysis in this table, the results for each farmer are weighted by the number of soybean base acres in the county.

ARC payments can change substantially (as shown earlier in figure 6 when using the deterministic framework), depending on the level of historic prices. The current high historic prices, and the fact that prices are declining (and potentially may remain substantially lower than historic prices), may play roles in which program farmers view as most beneficial for them over the next several years. Furthermore, producers may select both SCO and PLC for a particular crop and farm, but rules prevent a producer from selecting both ARC and PLC for a single crop on the same farm. While this is a consideration for producers to keep in mind, prices would have to drop substantially for PLC payments to come into play. For instance, soybean prices would have to drop below \$8.40 per bushel for a PLC payment to be made. If the realized price did drop below the PLC reference price of \$8.40 per bushel, a producer enrolled in SCO and PLC would then receive further benefits from the PLC program in addition to those of the SCO program (benefits that a grower enrolled in ARC would not receive). Producers will have to work through these issues when weighing the pros and cons of which programs to enroll in.

While prices can alter the level of support, the parameters of the various programs also matter. However, the program parameters will only change when policymakers alter the programs—typically in a Farm Act. The appendix contains an analysis of how the program parameters affect program efficacy in the section called "Looking ahead."

Mapping How Support Programs Affect Changes in Risk

The next part of this analysis explores how risk reduction at the county level varies across the United States. Corn, soybeans, and winter wheat have four maps each. The first map depicts the coefficient of variation of gross corn revenue, denoted as CV_0. The colors used for the map represent the level of risk as measured by the coefficient of variation. For example, dark blue counties have CV_0s below 0.35 and are the least risky with respect to corn revenue without support payments. The red counties have CV_0s above 0.55 and are the most risky counties with respect to corn revenue without support payments. The next three maps are the change in the coefficient of variation, denoted as ΔCV, for PLC, ARC, and SCO, where ΔCV is the CV of total gross revenue (gross revenue plus the support payment) less CV_0. For these maps, the program parameters (e.g., reference price, maximum coverage rate) are set at the actual values in the 2014 Farm Act. The green color corresponds to those counties where the ΔCV is greater than -0.01 and fall into the lowest risk reduction category. The blue counties have ΔCVs less than -0.03 and fall into the highest risk reduction category. On both the CV_0 and ΔCV maps, white counties indicates counties that we do not have continuous data for 1975 through 2013. As in the earlier section of the report, the simulations assume 2014 expected prices and yields.

In figure 12, a pattern emerges for the CV_0s of corn revenue. Most of the counties in Iowa and Illinois are low risk compared to the rest of the country. Indiana and southern Minnesota also contain a large portion of the corn-growing counties that have low revenue risk. Not surprisingly, these low-risk areas are where the majority of corn is grown in the United States, and the counties comprise the Corn Belt. Generally, as one moves farther away from the Corn Belt, corn revenue becomes more risky. There are high-risk counties for corn revenue throughout much of the United States. Such areas are in Texas, the Carolinas, and the Dakotas.

PLC, ARC, and SCO all have unique patterns of risk reduction across the United States. Figure 12b shows the risk reduction for corn under PLC. Throughout the country, with a few exceptions in Montana, Wisconsin, Pennsylvania, and New York, PLC provides very low risk reduction. The low revenue-risk reduction from PLC is not surprising since PLC is designed to reduce price risk, not revenue risk. Figure 12c shows that with corn ARC payments, most counties appear in the highest risk-reduction category and no counties fall in the lowest risk-reduction category. Counties in the middle revenue-risk reduction categories are scattered throughout the United States, with the exception of a cluster of the green and pink counties in southern Iowa and Illinois. This group of counties shows that the risk reduction from ARC payments has a weak negative correlation with CV_0. In other words, low revenue-risk counties may experience lower revenue-risk reduction compared to a high-risk area. Figure 12d demonstrates that many counties experience lower risk reduction from SCO compared to ARC, but the correlation between CV_0 and risk reduction can more easily be seen under SCO.

Figure 13a illustrates soybean revenue risk. The band of dark blue counties from eastern Nebraska through Indiana indicates low revenue risk for soybeans in this region. Emanating outward from these counties, soybean revenue tends to increase in riskiness. Soybean revenue in the Carolinas is at a higher risk compared to the majority of soybean-producing counties. High-risk counties can also be found in Arkansas and Mississippi near the Mississippi River. Figure 13b shows the effect of PLC on soybean revenue risk.

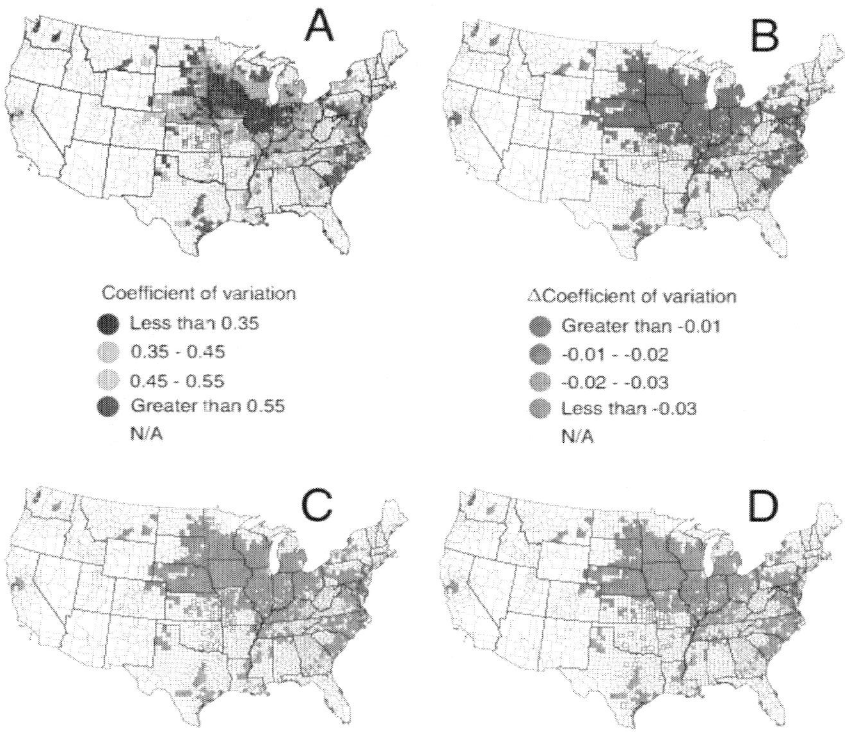

Source: USDA, Economic Research Service.

ARC = Agriculture Risk Coverage program; RP = Revenue Protection; SCO = Supplemental Coverage Option. N/A = data not available.

[1]The SCO calculation insures down to the 75-percent level, as if there were an underlying RP policy with 75-percent coverage. However, we assume there is no RP policy here in order to show the risk reduction impacts of SCO alone, so deeper losses otherwise covered by a 75-percent RP policy are realized by the producer in this scenario. Note that, in reality, SCO cannot be purchased independently of the underlying Federal crop insurance policy.

Figure 12. (A) Risk in corn gross revenue, measured as the coefficient of variation (CV) and changes in risk (ΔCV) for (B) PLC, (C) ARC, and (D) SCO1 where reference prices and coverage rates are held at the 2014 Farm Act levels.

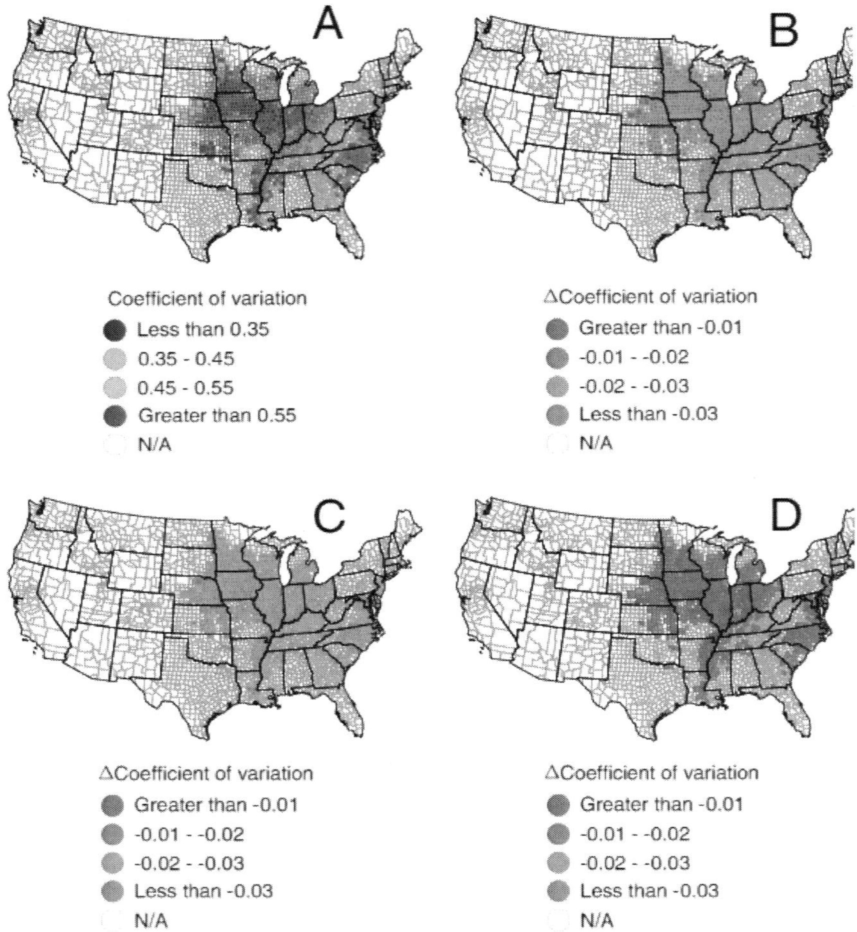

Source: USDA, Economic Research Service.

ARC = Agriculture Risk Coverage program; RP = Revenue Protection; SCO = Supplemental Coverage Option. N/A = data not available.

[1]The SCO calculation insures down to the 75-percent level, as if there were an underlying RP policy with 75-percent coverage. However, we assume there is no RP policy here in order to show the risk-reduction impacts of SCO alone, so deeper losses otherwise covered by a 75-percent RP policy are realized by the producer in this scenario. Note that, in reality, SCO cannot be purchased independently of the underlying Federal crop insurance policy.

Figure 13. (A) Risk in soybean gross revenue, measured as CV and changes in risk for (B) PLC, (C) ARC, and (D) SCO1 where reference prices and coverage rates are held at the 2014 Farm Act levels.

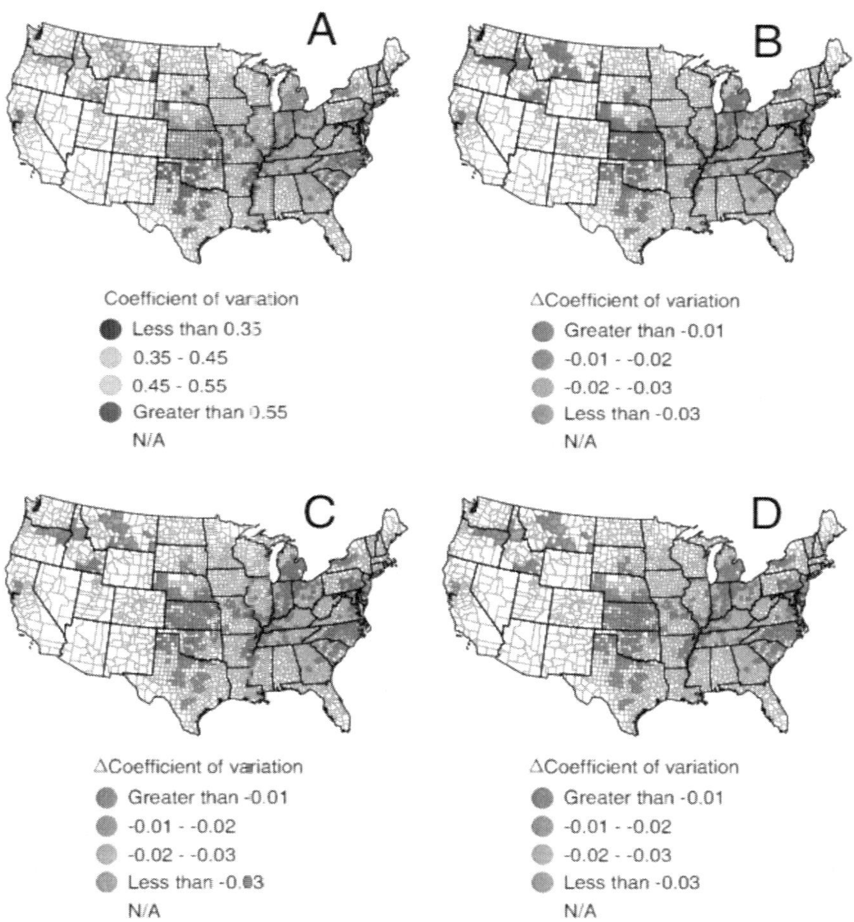

Source: USDA, Economic Research Service.

ARC = Agriculture Risk Coverage Program; RP = Revenue Protection; SCO = Supplemental Coverage Option. N/A = data not available.

[1]The SCO calculation insures down to the 75-percent level, as if there were an underlying RP policy with 75-percent coverage. However, we assume there is no RP policy here in order to show the risk reduction impacts of SCO alone, so deeper losses otherwise covered by a 75-percent RP policy are realized by the producer in this scenario. Note that, in reality, SCO cannot be purchased independently of the underlying Federal crop insurance policy.

Figure 14. (A) Risk in winter wheat gross revenue, measured as CV and changes in risk for (B) PLC, (C) ARC, and (D) SCO1 where reference prices and coverage rates are held at the 2014 Farm Act levels.

For soybean revenue, the risk reduction from PLC is minimal for all counties. Figure 13c and figure 13d illustrate that for soybeans, ARC and SCO tend to reduce more revenue risk compared to PLC, especially in high-risk counties. Like corn, ARC payments for soybean revenue tend to provide higher risk reduction compared to SCO.

Figure 14a illustrates that the production of winter wheat in the simulation is rather dispersed compared to the production of corn and soybeans. No wheat-producing counties fall into the lowest category of revenue risk. Pennsylvania and Michigan have comparatively low revenue risk for winter wheat, while winter wheat revenue in Oklahoma and Texas tends to be higher risk. Although figure 14b shows that the majority of counties experience very low revenue-risk reduction from PLC, some of the high revenue-risk counties in Texas and Oklahoma fall into the second lowest risk-reduction category. ARC and SCO do not present a clear pattern with regard to reducing revenue risk for winter wheat, which is made evident in figure 14c and figure 14d, respectively. Compared to ARC, SCO appears to provide higher risk reduction for winter wheat, contrary to the findings of soybeans and corn. This may help explain why wheat producers elected to cover 42 percent of wheat base acres with PLC, a much higher level than that found on corn or soybean acres (USDA/FSA, 2015).

CONCLUSION

PLC, which acts as a price floor, and SCO and ARC, colloquially known as "shallow-loss" programs, are new programs introduced in the 2014 Farm Act. These programs interact with each other and with the Federal crop insurance (FCI) program, generating a complex set of producer options and potential revenue outcomes. Each program affects agricultural risk exposure but does so in different ways. The underlying mechanics of these new programs and the ways in which the various available combinations of programs can impact crop revenue add to the knowledge base on the programs' impact on producer revenues and expected program costs.

Our analysis shows that mean payment and downside risk-reduction differences between SCO and ARC for soybean producers are quite sensitive to prices at planting time. From the standpoint of planting time in 2014, ARC would have provided a higher mean payment and generally a greater reduction in downside revenue risk than SCO, even if these differences were a relatively small portion of revenue. However, our analysis also shows that ARC payments in any given year are notably more sensitive than SCO payments to planting-time prices that year. Just as the producer's feeling about the future path of prices across the time span of the 2014 Farm Act may determine the choice between PLC and ARC (Effland et al., 2014), the decision between SCO and ARC may be also driven by the perception of future years' prices given that enrollment in ARC precludes enrolling in SCO for that crop. If the producers believe they face significant price risk, enrolling in SCO would also allow them to enroll in PLC, whereas enrolling in ARC would not permit that choice.

It is also possible that factors not directly related to returns per acre may tip the scales in favor of enrolling in either ARC or SCO. For example, an individual is limited to $125,000 in benefits generated from any combination of ARC payments, PLC payments, marketing loan gains, and/or loan deficiency payments (other than for peanuts, which has an additional, separate limit). While smaller farms may not need to worry about this limit, larger farms could run up against the imposed limits. On the other hand, enrolling in SCO requires the producer to pay a premium to enroll, even if the Government premium subsidy is 65 percent. The large number of options and how the programs interact with other programs complicate these decisions—particularly since these elections will last for the duration of the 2014 Farm Act.

Understanding how producers will respond to the new programs and how their decisions will be made will allow policymakers and other stakeholders to assess how the programs affect producer well-being. Modelling behavior often requires multiple assumptions that can be unrealistic and parameters that are often not known with certainty. Using portfolio analysis is also possible, but that relies on limited distributional information about producers' revenues. We therefore explore the concept of stochastic dominance in this chapter, which doesn't require unrealistic assumptions, while also taking into account the entire revenue distribution. However, this method often results in no clear choice for the producer.

Finally, the magnitude of risk reduction from these new Federal programs is dependent not only on the structure of the program, but also on the crop and the parameters of the program. For PLC, 2014 Farm Act reference prices lead to varied results among corn, soybeans, and winter wheat. In particular, for soybeans, the percentage of risk reduction is less than a tenth of risk reduction for corn and winter wheat. Also for corn and soybeans, PLC does not necessarily provide higher risk reduction for counties with high revenue risk. For ARC and SCO, as the coverage rate increases, the correlation between the standardized payment and the change in the coefficient of variation decreases, although this correlation remains strong. For corn and soybeans, ARC and SCO tend to have higher risk reduction for high-risk counties and lower risk reduction for low-risk counties, while the risk reduction among counties is more mixed for wheat. Corn and soybeans are grown in a relatively concentrated geographic area where the climate and soil conditions are relatively homogenous compared with where winter wheat is grown, which ranges from one side of the country to the other. As a result, the correlation between higher risk reduction in higher risk counties is stronger for corn and soybeans. In contrast, for winter wheat, idiosyncratic risk plays a larger role, likely due in part to geographic differences, generating a weaker link between the change in risk reduction and the level of risk faced by producers.

One caveat to these simulation results is that only counties with continuous production were included, which may have introduced a selection bias. Counties that sporadically grow a crop may be higher risk than the counties represented in the simulation. Therefore, high-risk counties may be underrepresented. In addition, the results can be sensitive to the assumption of the expected prices and yield used in the simulation, and, for ARC, the assumption of season-average prices in prior years that feed into the calculation of the revenue guarantee.

APPENDIX

Looking Ahead

Federal farm programs are set in stone for the life of the 2014 Farm Act—approximately 5 years. However, when Congress introduces a new Farm Bill, it could change the landscape of current Federal farm programs, including adding, eliminating, or making adjustments to current programs. In this appendix, we focus on the future, exploring how potential changes to key program parameters could alter the effectiveness of the Federal programs.

Program Parameters' Importance in Contributing to Revenue Risk Reduction

The simulation approach is used to examine the sensitivity of support to the program parameters. Key parameters affecting payments include the reference price for PLC (appendix table 1) and the coverage rate on benchmark or expected area revenue for ARC and SCO. Changes in these parameters have implications for Government costs, mean producer income, and reduction in the producer's revenue risk. For this chapter, we focus on the latter impact.

Appendix table 2 displays simulation results for the risk reduction of PLC using the actual reference prices as well as hypothetical reference prices both below and above the actual reference price to demonstrate the sensitivity to the choice of reference price. All else being equal, decreasing the reference price will lower the frequency with which payments are triggered and lower the size of the payments, thus providing lower revenue-risk reduction (and vice versa for increasing reference prices).

Appendix Table 1. Legislative prices in Title I of two Farm Acts, by covered commodities

Covered commodity	Unit	PLC reference price (2014 Farm Act)[1]	CCP target price (2008 Farm Act)[2]
Wheat	Bushel	5.50	4.17
Barley	Bushel	4.95	2.63
Oats	Bushel	2.40	1.79
Peanuts	Pound	0.2675	0.2475
Corn	Bushel	3.70	2.63
Grain sorghum	Bushel	3.95	2.63
Soybeans	Bushel	8.40	6
Dry peas	Pound	0.1100	0.0832
Lentils	Pound	0.1997	0.1281
Large chickpeas	Pound	0.2154	0.1281
Small chickpeas	Pound	0.2015	0.1036
Canola	Pound	0.2015	0.1268
Flaxseed	Bushel	11.2800	0.1268
Other oilseeds	Pound	0.2015	0.1268
Rice (long grain)	Pound	0.1400	0.105
Rice (medium grain)	Pound	0.1400	0.105
Upland cotton	Pound	NA	0.7125

[1] Source: 2014 Farm Act.

[2] Source: 2008 Farm Act. Table reflects final updates of target prices.

PLC = Price Loss Coverage program; CCP = Counter-cyclical Payment program; NA = not applicable—cotton is no longer considered a covered commodity under the 2014 Farm Act.

Appendix table 2 shows that PLC has the smallest risk reduction out of the three programs. The lowest reference price for corn causes changes in the CV less than 0.001, although increasing the reference price for corn from \$3.40 to \$4.00 changes the risk reduction from 0.14 percent to 3.55 percent. For soybeans, the reference prices of \$8.10 and \$8.40 cause changes in the CV for revenue no more than .001. Respectively, at these reference prices for soybeans, revenue risk decreases by 0.02 percent and 0.09 percent. Even at 30 cents higher than the reference price in the 2014 Farm Act, the risk reduction from PLC for soybeans is less than 1 percent. For winter wheat, the changes are moderate, with risk reduction ranging from 0.56 percent to 2.29 percent for the lowest to highest reference prices, respectively.

Appendix Table 2. Simulated national average of payments and change in variability of crop revenue under PLC, assuming alternative reference prices[1,2]

Corn				Soybeans				Winter wheat			
Reference price ($/bu)	Pay-ment ($/acre)	ΔCV	% ΔCV	Reference price ($/bu)	Payment	ΔCV	%ΔCV	Reference price ($/bu)	Payment ($/acre)	ΔCV	%ΔCV
$3.40	0.38	-0.0005	-0.14	$8.10	0.03	-0.0001	-0.02	$5.20	0.76	-0.0031	-0.56
$3.703	3.62	-0.0040	-1.18	$8.40	0. 17	-0.0003	-0.09	$5.50	1.70	-0.0066	-1.19
$4.00	12.16	-0.0122	-3.55	$8.70	0.54	-0.0010	-0.28	$5.80	3.37	-0.0124	-1.98

Source: USDA, Economic Research Service.

PLC = Price Loss Coverage program; bu = bushel; ΔCV is CV1-CV0, where CV0 is the coefficient of variation of gross crop revenue and CV1 is the coefficient of variation of total gross revenue (gross crop revenue plus the PLC payment). Coefficient of variation is the standard error of the revenue divided by its mean.

[1] Simulations assume 2014 expected price and yields.

[2] The results represent a weighted average of the revenues of a representative farmer in each county for which USDA's National Agricultural Statistics Service has reported yield data each year from 1975 through 2013, which is 1,001 counties for corn, 889 counties for soybeans, and 510 counties for winter wheat. In the data summaries, the results for each farmer are weighted by the number of base acres in the county. 3Actual reference price in the 2014 Farm Act.

In the 2014 Farm Act, the coverage rate for ARC is 86 percent of benchmark revenue. Appendix table 3 includes simulation results for not only this current coverage rate, but also coverage rates of 82 percent of benchmark revenue and 90 percent of benchmark revenue. The simulation results illustrate the risk reduction from ARC is much higher than the risk reduction from PLC, at least under the scenario of 2014 expected prices and yields. Corn experiences large risk reduction from ARC relative to soybeans and winter wheat. At coverage rates of 86 percent and 90 percent of benchmark revenue, risk reduction for corn is over 10 percent. Although the risk reduction is not as high for soybeans, the risk reduction for soybeans ranges from 5.36 percent to 7.95 percent for coverage rates of 82 percent to 90 percent of benchmark revenue. Winter wheat is a higher risk crop compared to corn and soybeans. Interestingly, the risk reduction from ARC is smaller for winter wheat compared to corn and soybeans. Even when looking at the ΔCV instead of percent-ΔCV, the changes in winter wheat are still smaller than the changes for corn and soybeans.

Appendix Table 3. Simulated national average of payments and change in variability of crop revenue under ARC, assuming alternative ARC coverage rates[1,2]

	Corn			Soybeans			Winter wheat		
Coverage rate	Payment	ΔCV	%ΔCV	Payment	ΔCV	%ΔCV	Payment	ΔCV	%ΔCV
82 percent	21.33	-0.031	-8.94	10.16	-0.020	-5.36	3.53	-0.017	-3.37
86 percent[3]	27.64	-0.035	-10.34	13.78	-0.025	-6.71	4.21	-0.020	-3.82
90 percent	34.23	-0.039	-11.45	17.78	-0.029	-7.95	4.97	-0.022	-4.29

Source: USDA, Economic Research Service.

ARC = Agriculture Risk Coverage program; ΔCV is CV_1-CV_0, where CV_0 is the coefficient of variation of gross crop revenue and CV_1 is the coefficient of variation of total gross revenue (gross crop revenue plus the Price Loss Coverage program payment). Coefficient of variation is the standard error of the revenue divided by its mean.

[1]Simulations assume 2014 expected price and yields.

[2]The results represent a weighted average of the revenues of a representative farmer in each county for which USDA's National Agricultural Statistics Service has reported yield data each year from 1975 through 2013, which is 1,001 counties for corn, 889 counties for soybeans, and 510 counties for winter wheat. In the data summaries, the results for each farmer are weighted by the number of planted acres in the county. [3]Actual ARC coverage rate in the 2014 Farm Act.

SCO currently allows for individuals to purchase coverage for up to 86 percent of the expected revenue/yield. Like the simulation results for ARC, the sensitivity analysis for SCO includes the current coverage rate of 86 percent as well as alternative coverage rates of 82 percent and 90 percent. Although producers need to have an underlying insurance policy to purchase SCO, indemnity payments from the underlying policy are not included here. By not including any Federal crop insurance indemnity payments, this allows for a more direct comparison between SCO and the other two programs.

Although ARC has higher payments than SCO and PLC for corn and soybeans, appendix table 3 illustrates that changing the coverage rate for SCO causes larger changes in the risk reduction compared to changing the coverage rate for ARC. For example in soybeans, changing the trigger from 82 percent of benchmark revenue to 90 percent of benchmark revenue for ARC results in an additional risk reduction of 2.59 percent.[16] However, for SCO, changing the coverage rate from 82 percent to 90 percent leads to an additional risk reduction of 4.52 percent. This larger increase in risk reduction for SCO holds for corn and winter wheat as well. The larger risk reduction in SCO stems from the harvest price protection built into revenue protection. Revenue protection (RP) is the underlying crop insurance policy used for the SCO simulations. The harvest-price protection calculates the revenue guarantee for SCO using the higher of the planting-time price or the harvest-time price. This flexibility in the guarantee leads to higher risk reduction. Producers of corn, soybeans, and wheat are able to enroll in PLC and SCO, potentially giving some additional downside risk protection over just SCO alone. Even under the scenario of 2014 expected prices and yields, in the case of corn, it appears that the risk reduction will be greater with the combination of PLC and SCO than with ARC; for soybeans, adding PLC provides little benefit, and for winter wheat, SCO alone already gives higher risk reduction benefits than ARC.

Relationships with Revenue Risk

Unlike the expired Direct Payments, payments from ARC, PLC, and SCO will tend to be counter-cyclical to crop revenue. That is, because PLC, ARC, and SCO payments are triggered when price or revenue fall below a threshold, these payments will tend to be made when farm income falls, thus partially compensating the farmer for the decrease in gross

---

[16] In actual practice, setting the SCO coverage rate at 85 percent or lower presumes that the coverage rate in the underlying traditional crop insurance program is less than the 85-percent coverage available for some crops in some regions.

revenue and reducing variability in farm revenue.[17] One way to assess the efficiency of a payment in targeting revenue risk is to examine the correlation of the payment with the change in revenue risk it provides. As revenue risk varies among crops and regions, we can also use correlation analysis to examine to what extent, if any, the programs' risk-reduction impacts target lower versus higher risk producers.

For each crop and program specification, to examine to what extent the payment is associated with revenue risk reduction, appendix tables 5 through 7 present the Pearson correlation coefficients[18] between the standardized payment and the change in CV of revenue (CV of total gross revenue minus CV of gross revenue) the payment provides. The standardized payment is the payment per acre divided by the mean gross crop revenue per acre; standardizing the payment puts it on a unit-free footing, simplifying its comparison to the CV by removing scale effects. If the correlation between the standardized payment and the change in CV is closer to -1, the payment is more closely associated with reducing revenue risk, e.g., the larger the payment, the more negative (decreasing) is the change in CV. Our expectation is that these correlations will be negative due to program design, and that the payments under a revenue program, such as ARC or SCO, will more closely target revenue risk than a program that targets only price, like PLC. However, a variety of factors could affect the measured correlations, including the level of reference prices relative to actual prices and empirical price-yield relationships.

[17] The SCO, ARC, and PLC program payments are not guaranteed to be countercyclical to farmer revenue; ARC and SCO are area payments, and, as such, have the possibility of being triggered even if an individual farmer revenue increases, or not being triggered when the farmer's revenue falls. Further, receiving an ARC or PLC payment does not mean a particular farmer's revenue has fallen, since these programs depend on a farmer's base, not what a farmer actually plants.

[18] The Pearson correlation coefficient is a statistic between -1 and 1. Negative 1 for the correlation coefficient indicates a perfectly linear negative relationship between two variables, while positive 1 indicates a perfectly linear positive relationship between two variables. When the correlation coefficient = 0, there is no linear relationship between the two variables.

To examine to what extent these payments target lower or higher risk producers, the tables also present the correlation of the change in CV of revenue with the CV of gross revenue. The closer this correlation is to zero, the more uniformly the program treats producers regardless of their revenue risk. The simulated averages of each of these variables for each county are used to calculate the correlation coefficient. In the interest of brevity, the change in the CV for crop revenue with and without support will be denoted as ΔCV and the CV for gross crop revenue (revenue without support payments) will be denoted as CV_0. As with appendix tables 2 through 4, appendix tables 5 through 7 show the results under actual reference prices and coverage rates and under the hypothetical lower and upper values.

For PLC, the correlation between the standardized payment and the change in revenue risk for each crop is negative in most cases, which can be seen in appendix table 5. The correlation coefficients for soybeans, corn, and winter wheat under the references prices in the 2014 Farm Act are -0.61, -0.78, and -0.96, respectively. The negative correlation indicates that when the standardized payment increases, the risk in the crop revenue is likely reduced. When the reference price for corn is $4.00 and $5.80 for winter wheat, the correlation between standardized payment and ΔCV is positive. This positive correlation indicates upside "risk" may be present. Therefore, total gross revenue, despite being higher with the PLC payment, becomes more volatile if the PLC payments are triggered by a relatively high reference price.

Under PLC, we see a negative relationship for corn and winter wheat between CV_0 and ΔCV, meaning that higher risk areas are likely to see higher decreases in revenue risk. For corn and winter wheat, as the reference price increases, this correlation becomes slightly stronger. For corn, the correlation between CV_0 and ΔCV is weak to moderate, spanning from -0.38 to -0.52 given the reference prices. Appendix table 5 also shows that the correlation for CV_0 and ΔCV is between -0.64 and -0.68 for winter wheat for the given reference prices. However, in the case of soybeans, the correlation between CV_0 and ΔCV is positive for all three reference prices. Although this relationship is fairly weak for soybeans, low-revenue-risk areas are more likely to experience larger decreases in revenue risk compared to higher risk areas.

Appendix table 6 illustrates that, as expected, ARC has more consistent patterns of correlation among the crops compared to PLC. The correlation between ΔCV and the standardized payment is the strongest for winter wheat and the weakest for corn. However, all three crops have strong negative correlation between ΔCV and the standardized payment for the coverage rates of 82 percent and 86 percent, with the correlation stronger than -0.85. As with PLC, strong negative correlation between the standardized payment and ΔCV indicates that increasing the support payments with respect to crop revenue will likely decrease the risk in crop revenue. For corn, soybeans, and winter wheat, there is a decrease in the correlation between ΔCV and the standardized payment when the coverage rate is 90 percent. When the coverage rate is 90 percent, the correlation between ΔCV and the standardized payment for corn, soybeans, and winter wheat is -0.50, -0.76, and -0.81, respectively. One possible explanation for the weakening correlation is the higher frequency of maximum payments at higher coverage rates. A higher frequency of maximum payments causes nonlinearity in the relationship between standardized payments and ΔCV, which weakens the linear relationship.

Despite the strong correlation between ΔCV and the standardized payment for ARC, the correlation between the ΔCV and CV_0 is somewhat weak. The correlation coefficient between ΔCV and CV_0 for corn is -0.40 at the 86-percent coverage rate and -0.43 for soybeans at the same rate. The correlation is weaker for winter wheat at -0.32 for the 86-percent coverage rate. One should note that, on average, the yield variation among farms is substantially higher for wheat compared to soybeans and corn. Therefore, winter wheat farm-level yields are not as closely tied to county yields compared to the yields of corn and soybeans. If farm yields are not closely tied to county yields, then risk reduction cannot be closely tied to CV_0 under a county-based revenue support program.

The correlation results for SCO for corn, soybean, and winter wheat producers, seen in appendix table 7, are more similar to ARC than PLC. The correlation between the standardized payment and ΔCV for SCO is strong and negative, and the correlation weakens as the maximum coverage increases. For example, with a coverage rate of 82 percent for SCO, the correlation between standardized payment and ΔCV for winter wheat is -0.96, but this correlation drops to -0.69 when the coverage rate for SCO is 90 percent. The cause for this drop in correlation is likely the same cause as the drop in correlation for the standardized payment and ΔCV seen in appendix table 6 for ARC. Increasing the coverage rate increases the frequency of maximum payments, which creates a nonlinearity in the trend between standardized payment and ΔCV.

Appendix Table 4. Simulated national average of payments and change in variability of crop revenue under SCO assuming alternative coverage rates[1,2,3]

Corn				Soybeans			Winter wheat		
Coverage rate	Payment	ΔCV	%ΔCV	Payment	ΔCV	%ΔCV	Payment	ΔCV	%ΔCV
82 percent	6.96	-0.015	-4.59	4.58	-0.012	-3.10	3.98	-0.021	-4.00
86 percent4	12.61	-0.025	-7.49	8.59	-0.021	-5.26	6.79	-0.033	-6.34
90 percent	19.80	-0.035	-10.54	13.79	-0.300	-7.62	10.01	-0.045	-8.69

Source: USDA, Economic Research Service.

SCO = Supplemental Coverage Option; ΔCV is CV1-CV0, where CV0 is the coefficient of variation of gross crop revenue and CV1 is the coefficient of variation of total gross revenue (gross crop revenue plus the Price Loss Coverage program payment). Coefficient of variation is the standard error of the revenue divided by its mean.

1 Simulations assume 2014 expected price and yields.

2 The results represent a weighted average of the revenues of a representative farmer in each county for which USDA's National Agricultural Statistics Service has reported yield data each year from 1975 through 2013, which is 1,001 counties for corn, 889 counties for soybeans, and 510 counties for winter wheat. In the data summaries, the results for each farmer are weighted by the number of planted acres in the county.

3 The SCO calculations assume that the underlying policy is Revenue Protection with 75-percent coverage rate. The SCO payments are net of the farmer-paid SCO premium.

4 Actual SCO coverage rate in the 2014 Farm Act.

The nonlinearity in turn decreases the correlation between the two variables. Unlike ARC, the correlation between CV_0 and ΔCV for SCO remains very similar across coverage rates. The correlation coefficients between CV_0 and ΔCV for SCO are -0.39, -0.45, and -0.21 for corn, soybeans, and winter wheat, respectively, at the coverage rates of 86 percent and 90 percent. Therefore, for SCO, increasing the coverage rate will not increase or decrease the ability to target risk reduction for high-risk areas.

Appendix Table 5. Correlations of the change in variability of revenue (ΔCV) with the payment and with the base variability of revenue: the case of PLC under alternative reference prices1,2,3

Corn			Soybeans			Winter wheat		
Reference price ($/bu)	Correlation of %ΔCV with:		Reference price ($/bu)	Correlation of %ΔCV with:		Reference price ($/bu)	Correlation of %ΔCV with:	
	Payment[e]	CV0		Payment[e]	CV0		Payment[e]	CV0
$3.40	-0.75	-0.38	$8.10	-0.59	0.38	$5.20	-0.95	-0.64
$3.704	-0.78	-0.45	$8.40	-0.61	0.36	$5.50	-0.95	-0.66
$4.00	0.14	-0.52	$8.70	-0.64	0.33	$5.80	0.35	-0.68

Source: USDA, Economic Research Service.

PLC = Price Loss Coverage program; bu = bushel.

ΔCV is CV1-CV0, where CV0 is the coefficient of variation of gross crop revenue and CV1 is the coefficient of variation of total gross revenue (gross crop revenue plus the PLC payment). Coefficient of variation is the standard error of the revenue divided by its mean.

[1] Simulations assume 2014 expected price and yields.

[2] The results represent a weighted average of the revenues of a representative farmer in each county for which USDA's National Agricultural Statistics Service has reported yield data each year from 1975 through 2013, which is 1,001 counties for corn, 889 counties for soybeans, and 510 counties for winter wheat. In the data summaries, the results for each farmer are weighted by the number of base acres in the county.

[3] The payment was standardized by dividing it by the mean gross crop revenue.

[4] Actual reference price in the 2014 Farm Act.

Appendix Table 6. Correlations change in variability of revenue (ΔCV) with the payment and with the base variability of revenue: the case of simulated ΔCV for ARC under alternative coverage rates[1,2,3]

	Corn		Soybeans		Winter wheat	
	Correlation of %ΔCV with:		Correlation of %ΔCV with:		Correlation of %ΔCV with:	
Coverage rate	Payment	CV_0	Payment	CV_0	Payment	CV_0
82 percent	-0.90	-0.42	-0.97	-0.46	-0.99	-0.34
86 percent[4]	-0.87	-0.40	-0.95	-0.42	-0.99	-0.32
90 percent	-0.50	-0.41	-0.75	-0.41	-0.81	-0.30

Source: USDA, Economic Research Service.

ARC = Agriculture Risk Coverage program.

ΔCV is CV_1-CV_0, where CV_0 is the coefficient of variation of gross crop revenue and CV_1 is the coefficient of variation of total gross revenue (gross crop revenue plus the Price Loss Coverage (PLC) payment). Coefficient of variation is the standard error of the revenue divided by its mean.

[1]Simulations assume 2014 expected price and yields.

[2]The results represent a weighted average of the revenues of a representative farmer in each county for which USDA's National Agricultural Statistics Service has reported yield data each year from 1975 through 2013, which is 1,001 counties for corn, 889 counties for soybeans, and 510 counties for winter wheat. In the data summaries, the results for each farmer are weighted by the number of base acres in the county.

[3]The payment was standardized by dividing it by the mean gross crop revenue.

[4]Actual coverage rate in the 2014 Farm Act.

Appendix Table 7. Correlations change in variability of revenue (ΔCV) with the payment and with the base variability of revenue: the case of SCO under alternative coverage rates[1,2,3]

	Corn		Soybeans		Winter wheat	
	Correlation of %ΔCV with:		Correlation of %ΔCV with:		Correlation of %ΔCV with:	
Coverage rate	Payment	CV_0	Payment	CV_0	Payment	CV_0
82 percent	-0.90	-0.39	-0.96	-0.46	-0.96	-0.22
86 percent[4]	-0.87	-0.39	-0.94	-0.45	-0.94	-0.21
90 percent	-0.63	-0.39	-0.78	-0.45	-0.69	-0.21

Source: USDA, Economic Research Service.

SCO = Supplemental Coverage Option; ΔCV is CV_1-CV_0, where CV_0 is the coefficient of variation of gross crop revenue and CV_1 is the coefficient of variation of total gross revenue (gross crop revenue plus the Price Loss Coverage (PLC) payment). Coefficient of variation is the standard error of the revenue divided by its mean.

[1]Simulations assume 2014 expected price and yields.

[2]The results represent a weighted average of the revenues of a representative farmer in each county for which USDA's National Agricultural Statistics Service has reported yield data each year from 1975 through 2013, which is 1,001 counties for corn, 889 counties for soybeans, and 510 counties for winter wheat. In the data summaries, the results for each farmer are weighted by the number of base acres in the county.

[3]The payment was standardized by dividing it by the mean gross crop revenue.

[4]Actual coverage rate in the 2014 Farm Act.

Appendix Table 8. Summary of program payments and total revenue producer ($/acre), soybeans

a. All counties						
	Not enrolled[1]	**SCO only**[2]	**Program enrollment**		**RP+SCO+PLC**	**RP+ARC**[3]
			ARC only[3]	**RP (75%)**		
Average (net) payment	—	9	14	19	27	31
Average total revenue	498	507	520[2]	517	533	538[2]
Coefficient of variation of revenue[4]	0.40	0.38	0.36	0.30	0.28	0.27
95-percent confidence interval of revenue	[151, 910]	[177, 909]	[195, 912]	[359, 894]	[362, 895][5]	[367, 897][5]
b. Low-risk county (Linn County, IA)						
Program enrollment						
	Unenrolled	SCO only[2]	ARC only	RP (75%)	RP+SCO+PLC	RP+ARC
Average payment (net of premium)	—	8	13	9	17	22
Average total revenue	570	578	583	579	587	592
Coefficient of variation of revenue[4]	0.28	0.26	0.26	0.24	0.22	0.22
95-percent confidence interval	[287, 911]	[315, 909]	[319, 911]	[417, 904]	[413, 902][5]	[417, 904][5]
c. Medium-risk county (Stearns County, MN)						
Program enrollment						
	Not enrolled	SCO only[2]	ARC only	RP (75%)	RP+SCO+PLC	RP+ARC
Average (net) payment	—	10	14	33	42	47
Average total revenue	457	466	471	489	499	503
Coefficient of variation of revenue[4]	0.55	0.53	0.52	0.39	0.37	0.36
95-percent confidence interval	[0, 993]	[42, 993]	[42, 994]	[317, 967]	[312, 966][5]	[317, 968][5]
d. High-risk county (Prentiss County, MS)						
Program enrollment						
	Not enrolled	SCO only[2]	ARC only	RP (75%)	RP+SCO+PLC	RP+ARC
Average (net) payment	—	11	15	41	52	56
Average total revenue	324	335	339	365	376	380

d. High-risk county (Prentiss County, MS)						
	Not enrolled[1]	SCO only[2]	Program enrollment ARC only[3]	RP (75%)	RP+SCO+PLC	RP+ARC[3]
Coefficient of variation of revenue[4]	0.76	0.73	0.72	0.49	0.47	0.47
95-percent confidence interval	[0, 871]	[-6, 877]	[0, 879]	[223, 837]	[217, 843][5]	[223, 845][5]

Source: USDA, Economic Research Service.

Note: Payments and total revenue (gross revenue plus payments) are net of the producer-paid premiums for Revenue Protection and Supplemental Coverage Option.

ARC = Agriculture Risk Coverage program; PLC = Price Loss Coverage program; RP = Revenue Protection; SCO = Supplemental Coverage Option. [1]For the summary across all counties, this column "not enrolled" shows statistics for gross revenue where the weighted average across counties is weighted by acres planted to soybeans in the county. However, for the simulations involving ARC, soybean base acres per county is used as the weight. In that case, the average gross revenue, coefficient of variation (CV), and 95-percent confidence interval is $506, 0.38, and [$166, $910], respectively. [2]"SCO only" is not available in practice but is used to allow a direct comparison with ARC. However, it still requires a lower bound on the coverage (since SCO covers down to the underlying policy in place). We therefore impose a lower limit of 75 percent (as if there was an RP policy in place that had a coverage rate of 75 percent. [3]With ARC, the results across counties are weighted by base acres, not planted acres. In this case, the mean and 95-percent confidence interval of "unenrolled" (i.e., gross revenue) are $507/acre and [$166/acre, $911/acre], respectively, and not those in the second column of the table. This confidence interval is nonparametric (i.e., does not assume any particular distribution), and as such, may not be symmetric around the average. [4]"Coefficient of variation of revenue" is the standard deviation of revenue divided by average revenue, and is a standardized measure to allow comparability of variability across programs. [5]While these are likely close to the true interval, we do not believe them—they are the result of averaging across counties while weighting by base acres. Adding ARC (or SCO plus PLC) to RP will not cause the lower end of the distribution to change for any individual county (see fig. 11). It will shift many producers to higher levels of revenues, indicated by the increase in the average total revenue, and will lower the CV of revenue, but will not shift the lower end of the distribution because these are area-based policies/programs and, as a result, will not provide benefits for everyone suffering losses—and there will be enough who do not correlate perfectly with the county outcomes such that the lower bound of the 95-percent confidence interval will remain unchanged. With respect to the upper bound, it is not clear how RP+SCO+PLC can have a higher upper bound than RP alone, given SCO also has a premium associated with it, which should lower that upper bound.

Appendix Table 9. Sensitivity of mean ARC and net SCO payments—with and without gross revenue—to expected prices (E[P]) being 50 percent and 150 percent of the 2014 expected price, soybeans

	ARC		**Net SCO1(assuming underlying 75-percent coverage revenue policy in place)**	
	Lower E[P] $5.68/bu	Higher E[P] $17.04/bu	Lower E[P] $5.68/bu	Higher E[P] $17.04/bu
Mean payments per acre				
All U.S. counties2	41 (8 ; 47)	1 (0 ; 10)	4 (-2 ; 29)	13 (-7 ; 86)
Low-risk county (Linn Cty, IA)	47 (3 ; 52)	0 (0 ; 0)	4 (0 ; 35)	12 (-6 ; 105)
Medium-risk county (Stearns Cty, MN)	36 (0 ; 42)	1 (0 ; 27)	5 (-3 ; 28)	14 (-8 ; 84)
High-risk county (Prentiss Cty, MS)	29 (0 ; 31)	3 (0 ; 31)	5 (-3 ; 20)	16 (-9 ; 61)
Gross revenue per acre				
All U.S. counties1,3	253 (83 ; 455)	760 (249 ; 1366)	249 (76 ; 455)	748 (227 ; 1365)
Low-risk county (Linn Cty, IA)	285 (144 ; 456)	855 (430 ; 1367)	285 (144 ; 456)	855 (430 ; 1367)
Mean payments per acre				
Medium-risk county (Stearns Cty, MN)	228 (0 ; 497)	685 (0 ; 1490)	228 (0 ; 497)	685 (0 ; 1490)
High-risk county (Prentiss Cty, MS)	162 (0 ; 435)	489 (0 ; 1308)	162 (0 ; 435)	487 (0 ; 1308)

	ARC		Net SCO1(assuming underlying 75-percent coverage revenue policy in place)	
	Lower E[P] $5.68/bu	Higher E[P] $17.04/bu	Lower E[P] $5.68/bu	Higher E[P] $17.04/bu
Gross revenue plus payment per acre				
All U.S. counties1	294 (128 ; 493)	761 (253 ; 1366)	253 (89 ; 454)	760 (266 ; 1363)
Low-risk county (Linn Cty, IA)	331 (194 ; 502)	855 (430 ; 1367)	289 (158 ; 455)	866 (473 ; 1364)
Medium-risk county (Stearns Cty, MN)	264 (42 ; 527)	686 (0 ; 1490)	233 (21 ; 496)	699 (63 ; 1489)
High-risk county (Prentiss Cty, MS)	191 (31 ; 465)	489 (0 ; 1308)	168 (-3 ; 438)	503 (-9 ; 1317)

Source: USDA, Economic Research Service.

Note: Actual expected price = $11.36/bu at time of planting in 2014; $5.68 = 0.5*$11.36 and $17.04 = 1.5*$11.36 are used to explore how variations in expected prices can alter program support. ARC = Agriculture Risk Coverage program; SCO = Supplemental Coverage Option; bu = bushel. Numbers in parentheses are lower and upper bounds of a 95-percent confidence interval. 1"Net SCO" is the SCO payment less what the farmer paid for SCO policy. 2The statistics for the gross revenue distributions for "all U.S. counties" do not exactly match across the ARC and SCO scenarios, as the former is weighted by base acres and the latter by planted acres. 3"All U.S. counties" represents the analysis of a representative farmer in all counties for which USDA's National Agricultural Statistics Service has reported soybean yield data each year from 1975 through 2013, which is 889 counties. In the data summaries, the results for each farmer are weighted by the number of planted or base acres in the county.

Appendix Table 10. How ARC and SCO change the expected revenue's lower bound of the 95-percent confidence interval (for all counties, soybeans)

Lower bound of revenue 95-percent confidence interval ($/acre)			
Fraction of price	**Gross revenue per acre**	**Gross revenue plus program payment per acre**	**Difference ($/acre)**
	Agriculture Risk Coverage (ARC)		
0.5*E[P]	83	125	42
E[P]	166	194	28
1.5*E[P]	249	253	4
	Supplemental Coverage Option (SCO)		
0.5*E[P]	76	89	13
E[P]	151	177	26
1.5*E[P]	227	266	39

Source: USDA, Economic Research Service.

ARC = Agriculture Risk Coverage program; SCO = Supplemental Coverage Option.

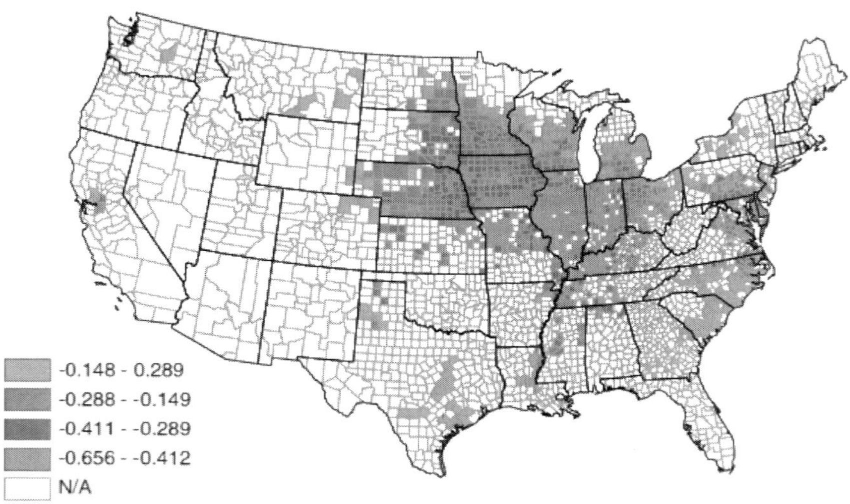

Source: USDA, Economic Research Service.

N/A = data not available.

Appendix Figure 1. Correlation between the yields of each county's representative farm and the national price for corn.

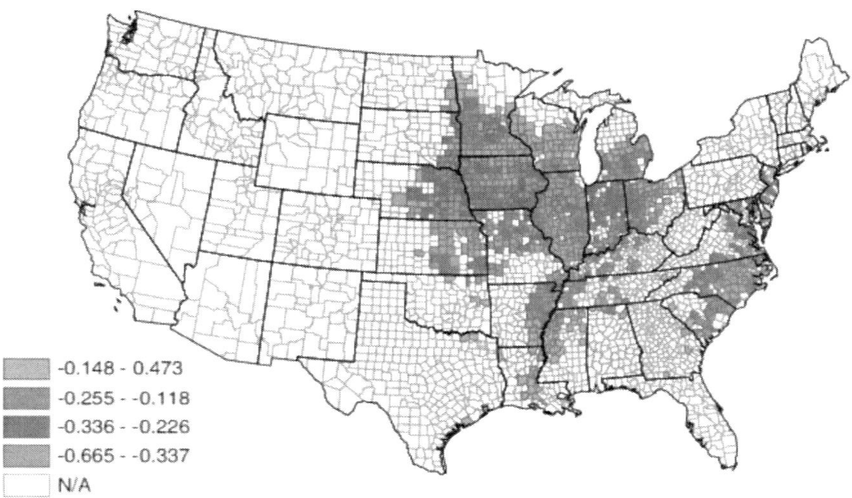

Source: USDA, Economic Research Service.
N/A = data not available.

Appendix Figure 2. Correlation between the yields of each county's representative farm and the national price for soybeans.

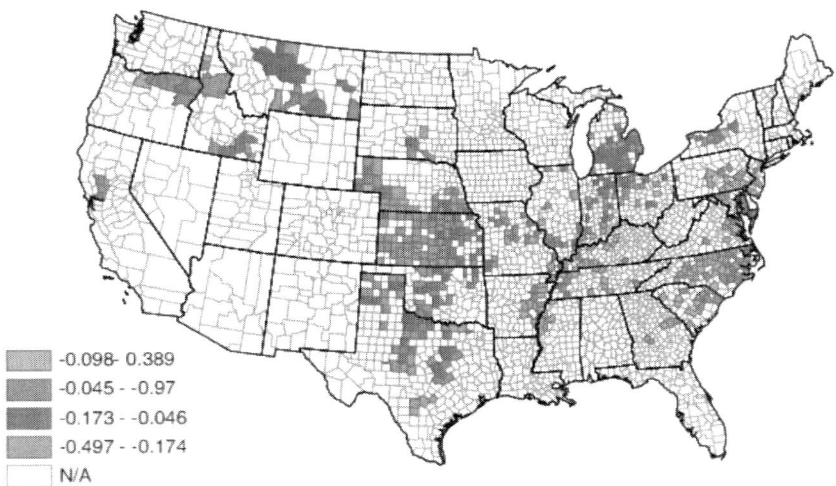

Source: USDA, Economic Research Service.
N/A = data not available.

Appendix Figure 3. Correlation between the yields of each county's representative farm and the national price for winter wheat.

References

Arrow, K. 1963. "Uncertainty and the Welfare Economics of Medical Care," *The American Economic Review*, 8(5):941-973.

Coble, K., and R. Dismukes 2008. "Distributional and Risk Reduction Effects of Commodity Revenue Program Design," *Review of Agricultural Economics* 30:543-553.

Cooper, J., and B. Delbecq, 2014a. *A Multi-Region Approach to Assessing Fiscal and Farm Level.*

Consequences of Government Support for Farm Risk Management," paper presented at *AIEAA Conference*, June 2014.

Cooper, J., and B. Delbecq, 2014b. "A Multi-Region Approach to Assessing Fiscal and Farm Level Consequences of Government Support for Farm Risk Management," *Bio-Based and Applied Economics* 3(3):205-227.

Cooper, J. 2010. "Average Crop Revenue Election: A Revenue-Based Alternative to Price-Based Commodity Payment Programs," *American Journal of Agricultural Economics* 92(4):1214-1228.

Cooper, J. 2009. *Economic Aspects of Revenue-Based Commodity Support*, USDA, Economic Research Report, ERR-72. http://www.ers.usda.gov/ publications/err-economic-research-report/ err72.aspx.

Effland, Anne, Joseph Cooper, and Erik O'Donoghue. 2014. "2014 Farm Act Shifts Crop Commodity Programs Away From Fixed Payments and Expands Program Choices," *Amber Waves*, July.

Harwood, J., R. Heifner, K.H. Coble, J. Perry, and A. Somwaru. 1999. *Managing Risk in Farming: Concepts, Research, and Analysis*, USDA, Economic Research Service, AER-774. http://www. ers.usda.gov/ publications/aer-agricultural-economic-report/aer774.aspx

Heimlich, R. 2000. *Farm Resource Regions*, USDA, Economic Research Service, AIB-760. http:// www.ers.usda.gov/publications/aib-agricultural-information-bulletin/aib760.aspx.

Levy, H. 1992. "Stochastic Dominance and Expected Utility: Survey and Analysis," *Management Science* 38(4):555-593.

Miranda, M.J., and J.W. Glauber. 1997. "Systemic Risk, Reinsurance, and the Failure of Crop Insurance Markets." *American Journal of Agricultural Economics* 79(1):206-215.

Miranda, M.J. 1991. "Area-Yield Insurance Reconsidered." *American Journal of Agricultural Economics* 73(2).233-242.

Serra, T., D. Zilberman, B. Goodwin, and A. Featherstone. 2006. "Effects of decoupling on the mean and variability of output," *European Review of Agricultural Economics* 33(3):269-288.

U.S. Department of Agriculture, Economic Research Service. 2014a. *Risk Management Maps and Charts.* Dec. 3, 2014. http://www.ers.usda.gov/topics/farm-practices-management/risk-management/maps-charts.aspx.

U.S. Department of Agriculture, Economic Research Service. 2014b. *Agricultural Act of 2014: Highlights and Implications*. http://www.ers.usda.gov/agricultural-act-of-2014-highlights-andimplications.aspx.

U.S. Department of Agriculture, Farm Service Agency. 2014. "Base Acre Reallocation, Yield Updates, Agriculture Risk Coverage (ARC) & Price Loss Coverage (PLC)," *2014 Farm Bill Fact Sheet*, September 2014. http://www.fsa.usda.gov/Internet/FSA_File/base_acre_reallo cate_arc_ plc.pdf.

U.S. Department of Agriculture, Farm Service Agency. 2015. *ARC/PLC Program. June 2015.* http://www.fsa.usda.gov/programs-and-services/arcplc_program/index

U.S. Department of Agriculture, National Agricultural Statistics Service. *Quick Stats*, various years. http://quickstats.nass.usda.gov/

U.S. Department of Agriculture and the National Association for Agriculture and Food Policy. 2014. *USDA/NAAFP Decision Aids.* https://usda.afpc.tamu.edu/

U.S. Department of Agriculture, Risk Management Agency, Federal Crop Insurance Corporation. 2014. *Supplemental Coverage Option Endorsement*, 15-SCO. http://www.rma.usda.gov/policies/2015/15sco.pdf

U.S. Department of Agriculture, Risk Management Agency. 2014a. *Summary of Business Reports and Data: National Summary by*

Insurance Plan—2014 crop year. http://www3.rma.usda.gov/ apps/sob/current_week/insplan2014.pdf

U.S. Department of Agriculture, Risk Management Agency. 2014b. *Supplemental Coverage Option for Federal Crop Insurance* fact sheet, October 2014. http://www.rma.usda.gov/news/currentissues/farmbill/2014NationalSupplementalCoverageOption.pdf

U.S. Department of Agriculture, Risk Management Agency. 2014c. *Soybeans–Wisconsin* fact sheet. August 2014. http://www.rma.usda.gov/fields/mn_rso/2014/2014wisoybeans.pdf

U.S. Department of Agriculture, Risk Management Agency. 2014d. *Daily Prices*. http://www3.rma. usda.gov/apps/pricediscoveryweb/ Daily Prices.aspx

Wøodzimierz, O., and A. Ruszczy. 1999. "From stochastic dominance to mean-risk models: Semi-deviations as risk measures," *European Journal of Operational Research* 116(1999):33-50.

Xiaohui, D., B. Barnett, and D. Vedenov. 2007. "Is There a Viable Market for Area-Based Crop Insurance?" *American Journal of Agricultural Economics* 89(2) (May):508–519.

In: Agricultural Issues
Editor: Michael Butlin

ISBN: 978-1-53615-474-0

Chapter 8

FEDERAL CROP INSURANCE: SPECIALTY CROPS*

Isabel Rosa and Renée Johnson

SUMMARY

The federal crop insurance program offers subsidized crop insurance policies to farmers. Farmers can purchase policies that pay indemnities when their yields or revenues fall below guaranteed levels. While the majority of federal crop insurance policies cover yield or revenue losses, the program also offers policies with other types of guarantees, such as index policies that trigger an indemnity payment based on weather conditions. The Federal Crop Insurance Corporation (FCIC), a government corporation within the U.S. Department of Agriculture (USDA), pays part of the premium—about 63%, on average—across the federal crop insurance portfolio during crop year 2017, while policy holders—farmers and ranchers—pay the balance. Private insurance companies, known as Approved Insurance Providers (AIPs), deliver the policies in return for administrative and operating subsidies from FCIC. AIPs also share underwriting risk with FCIC through a mutually

* This is an edited, reformatted and augmented version of Congressional Research Service, Publication No. R45459, dated January 14, 2019.

negotiated Standard Reinsurance Agreement. The USDA Risk Management Agency (RMA) administers the federal crop insurance program.

The federal crop insurance program primarily covers traditional field crops (such as wheat, corn, and soybeans) that are supported by USDA's revenue-support programs. Unlike these traditional crops, *specialty crops*—defined in statute as "fruits and vegetables, tree nuts, dried fruits, and horticulture and nursery crops (including floriculture)" (7 U.S.C. §1621 note) — have not been a major part of federal crop insurance support. Specialty crops are also generally not eligible for USDA's revenue-support programs. USDA estimates that the statutory definition of *specialty crops* covers more than 300 agricultural commodities, including fresh and processed fruits and vegetables, tree nuts, nursery plants (trees, shrubs, and flowering plants), herbs, spices, coffee, tea, honey, and maple syrup.

Legislative changes, coupled with ongoing administrative efforts by USDA, have expanded federal crop insurance coverage for specialty crops, and they now account for a small but growing number of federal crop insurance policies bought by farmers. Among the issues Congress may consider if it seeks to further expand coverage for specialty crops are data collection and price discovery for commodities not sold on exchanges (such as most fruits and vegetables), coverage of quality losses, and the effect of ad hoc payments on the demand for crop insurance.

Federal crop insurance policies currently cover around 38 specialty crop categories, which include roughly 80 types of fruits, vegetables, tree nuts, and nursery crops. Over the past few decades, total specialty crop insured liabilities rose from nearly $1 billion in 1989 to nearly $18.5 billion in 2017. In 2017, federal crop insurance policies covered about 9 million acres of specialty crops, around 800,000 bee colonies, about 100,000 tons of raisins, and roughly 60,000 fruit and coffee trees. Across all specialty crops, coverage and the premium subsidy paid by the federal government may vary depending on the crop. Moreover, for many specialty crops, crop-specific insurance policies are not available. Currently, about one-half of all U.S. specialty crop acres are covered by federal crop insurance policies. Some specialty crops may be covered under a Whole Farm Revenue Protection (WFRP) insurance policy. WFRP is designed to fill in coverage gaps for producers of uninsured crops that lack individual policy coverage and for producers marketing to local, farm-identity preserved, or direct markets. The average premium subsidy rate for WRFP was about 70% in 2017. Federal crop insurance for specialty crops and WFRP together accounted for about 17% of the entire federal crop insurance portfolio, as measured by liability, during crop year 2017.

INTRODUCTION

The federal crop insurance program offers subsidized crop insurance policies to farmers. Historically, the federal crop insurance program has covered primarily traditional field crops such as wheat, corn, and soybeans. In contrast, *specialty crops*—covering fruits, vegetables, tree nuts, and nursery crops—have not been a major part of the federal crop insurance program. However, legislative changes, coupled with ongoing administrative efforts by the U.S. Department of Agriculture (USDA), have expanded federal crop insurance coverage for specialty crops, and they now account for a small but growing number of federal crop insurance policies bought by farmers. Over the past few decades, total specialty crop insured liabilities rose from nearly $1 billion in 1989 to nearly $18.5 billion in 2017. Federal crop insurance policies currently cover around 38 specialty crop categories, which include roughly 80 types of fruits, vegetables, tree nuts, and nursery crops.

Many specialty crops, however, do not have crop-specific insurance policies. Currently, about one-half of all U.S. specialty crop acres are covered by federal crop insurance policies. Some specialty crops may be covered under a Whole Farm Revenue Protection (WFRP) insurance policy, intended to fill in coverage gaps for producers of uninsured crops that lack individual policy coverage and for producers marketing to local, farm-identity preserved, or direct markets. Despite this expansion, coverage for specialty crops remains below that for traditional crops. Combined, federal crop insurance for specialty crops and WFRP together accounted for about 17% of the entire federal crop insurance portfolio by liability during crop year 2017.

This chapter focuses on how specialty crops are covered under the federal crop insurance program.

Crop Insurance Program Authority and Operation

Overview of the Federal Crop Insurance Program

The federal crop insurance program is permanently authorized by the Federal Crop Insurance Act, as amended (7 U.S.C. §1501 *et seq.*). The USDA's Risk Management Agency (RMA) regulates the program, and the Federal Crop Insurance Corporation (FCIC) funds it.

Congress first authorized federal crop insurance in an effort to mitigate the effects on farmers of the Great Depression and of the crop losses seen in the Dust Bowl. In 1938, the FCIC was created to carry out the program, which focused on major crops like wheat in leading producing regions. During the same era, farm programs were established to support crop prices and boost farm income for producers of so-called program crops, including wheat, corn, and cotton.

The availability of federal crop insurance expanded with the passage of the Federal Crop Insurance Act of 1980 (P.L. 96-365), which brought coverage to many more crops and regions of the country. To increase participation, Congress enhanced the crop insurance program in 1994 by raising subsidy levels and in 2000 by expanding geographic availability and again raising subsidy levels.[1] The changes also expanded the role of the private sector in developing new products that would help farmers manage their risks. Today, many banks, when making operating loans, require that farmers purchase crop insurance.

The federal crop insurance program provides farmers with risk management tools to cope with yield and price declines. Under the program, farmers can purchase subsidized policies that pay indemnities when their production or revenue[2] falls below a producer-selected coverage

---

[1] For more on the history of federal crop insurance, see RMA, "History of the Crop Insurance Program," hhttps://www.rma.usda.gov/About-RMA/History-of-RMA; and CRS Report R45193, *Federal Crop Insurance: Program Overview for the 115th Congress*.

[2] The majority of federal crop insurance policies cover yield or revenue losses, but the program also offers policies with other types of guarantees, such as index policies that trigger indemnity payments based on weather conditions.

level. Insurance policies are sold and completely serviced through approved private insurance companies. The insurance companies' losses are reinsured by USDA, and their administrative and operating costs are subsidized by the federal government.

In purchasing a crop insurance policy, a producer selects a level of coverage (i.e., deductible) and pays a portion of the premium that increases with higher coverage levels (or none of it in the case of catastrophic coverage). The federal government pays the rest of the premium (63%, on average, in 2017).

The federal crop insurance program offers two main levels of coverage: Catastrophic Risk Protection (CAT) and buy-up coverage:[3]

1) **CAT coverage** insures against losses in excess of 50% of normal yield, equal to 55% of the estimated market price of the crop (called 50/55 coverage). The premium is 100% subsidized. Producers pay an administrative fee of $655 per crop per county.[4] Limited-resource producers may have this fee waived.[5] CAT coverage is not available on all types of policies.
2) **Buy-up coverage** allows producers to obtain coverage beyond the catastrophic level and pay a premium that is subsidized by the federal government. Buy-up provides for additional coverage up to 85% of production per acre and 100% of a specified market price established for each crop and region. The premium subsidy for these policies ranges from 38% to 67%.[6] Producers pay an administrative fee of $30 per crop per county.[7]

---

[3] For additional background, see RMA, "Insurance Plans," https://www.rma.usda.gov/policies.

[4] 7 U.S.C. 1508 §(b)(5)(A).

[5] Information about how USDA defines *limited-resource producers* is available at https://lrftool.sc.egov.usda.gov/.

[6] The premium subsidy rate declines with higher coverage levels. For example, the subsidy rate is 67% for 50% yield coverage with 55% price coverage (i.e., yield loss must greater than 50%, and more than 55% of expected market price) but declines to 38% for yield coverage of 80% with 100% price coverage (i.e., yield loss must greater than 20%).

[7] 7 U.S.C. 1508 §(c)(10)(A).

The availability of crop insurance for a particular crop in a particular region is an administrative decision made by RMA. The decision is made on a crop-by-crop and county-by-county basis based on farmer demand for coverage and the level of risk associated with the crop in the region, among other factors. In areas where a policy is not available, farmers may request that RMA expand the program to their county. The process usually starts with a pilot program in order for RMA to gain experience and test the program components before it becomes more widely available. Alternatively, a policy can be reviewed and later discontinued if it fails to perform at an acceptable level (e.g., it experiences low participation or high losses).

RMA also regularly responds to requests from commodity organizations or industry representatives for enhancements to existing coverage, such as adding revenue coverage. RMA offers pilot programs with various types of coverage for new crops (particularly specialty crops) or that include new geographical areas. It uses the performance of these programs to inform its decision on whether to extend coverage permanently.[8]

Expansion of Coverage for Specialty Crops

Efforts to expand federal crop insurance coverage to crops that had not traditionally been eligible under the program dates back to the 1990s. The Federal Crop Insurance Reform and Department of Agriculture Reauthorization Act of 1994 (P.L. 103-354) helped initiate efforts to expand federal crop insurance coverage for specialty crop producers. It amended the Federal Crop Insurance Act to require (among other things) additional data collection, reporting to facilitate the development of new crop insurance policies, tracking the progress of newly developed policies, and an expected timetable for expanding crop insurance coverage to

[8] The FCIC Board of Directors determines when a policy goes from pilot to permanent program status. See RMA, "RMA Permanent and Pilot Program Designations," https://legacy.rma.usda.gov/policies/2017/permanent-pilot.pdf.

include "new and specialty crops."[9] Since then RMA has submitted episodic reports to Congress tracking this progress.[10] The Agriculture Risk Protection Act of 2000 (P.L. 106-224) further emphasized increasing the availability of risk management tools for producers of crops with no individual policy coverage.

Previous omnibus farm bills—including the Agricultural Act of 2014 (P.L. 113-79, "2014 farm bill") and the Food, Conservation, and Energy Act of 2008 (P.L. 110-246, "2008 farm bill")—also enhanced the federal crop insurance program by expanding its scope and broadening policy coverage for specialty crops, organic agriculture, and crops that are sold directly to local markets. These laws authorized and extended other USDA programs that help facilitate the formulation of crop insurance policies for specialty crop producers, including market data collection programs.

The 2008 and 2014 farm bills further expanded provisions governing how RMA administers and implements Section 508(h) of the Federal Crop Insurance Act, which governs new policy development, including how it is contracted out and funded, how policy ratings are undertaken, and how a policy may start as a pilot and may (or may not) evolve to a full-fledged insurance policy. Additional information on the Section 508(h) process is described in "Development of New Policies and Section 508(h)."

The 2014 farm bill also directed RMA to develop the WFRP insurance policy as part of an effort to provide additional options for crops that lack crop-specific policy coverage. WFRP policies insure revenue of the entire farm, rather than an individual crop, under a single insurance policy that is not limited to specialty crop production. The types of producers eligible for whole farm policies include direct-to-consumer marketers and producers of multiple agricultural commodities, including specialty crops, industrial

---

[9] See, for example, P.L. 103-354, §§105-106.

[10] These reports are required under Section 508(a)(6)(B) of Federal Crop Insurance Act, as amended, tracking the progress and expected timetable for expanding crop insurance coverage to new and specialty crops (7 U.S.C. §1508). Previous RMA reports include *1999 Report to Congress: New and Specialty Crops* (no date), *2001 Report to Congress: New and Specialty Crops* (no date), *Report to Congress: Specialty Crop Report* (November 2010), and *Report to Congress: Specialty Crop Report* (April 2015). Reports cover the 1999, 2001, 2009, and 2014 crop years. Annual data are also available from RMA at https://prodwebnlb.rma.usda.gov/apps/SummaryofBusiness/ReportGenerator.

crops, livestock, and aquaculture products. Coverage is also expanded for the value of packing, packaging, or any other similar on-farm activity. WFRP can be combined with other policies with buy-up coverage. WFRP does not offer CAT-level coverage and cannot be combined with CAT-level policies. WFRP is discussed further in "Whole Farm Revenue Protection (WFRP) Insurance."

Despite expansion of the federal crop insurance program, coverage for specialty crops remains below that for traditional crops in terms of individual crop policies and covered acres. Often insurers face technical difficulties in developing new policies such as price discovery, nonweather risks, and premium ratings. In some cases, USDA has not pursued policies for particular commodities because producers have expressed concerns that offering insurance could adversely affect the market (i.e., because an insurance policy reduces producer risk, farmers may plant more acreage, which could drive down prices and total crop revenue). This has been a particular concern for vegetable crops and explains in part lower levels of insured vegetable acreage compared with other crops.[11] These and other challenges are discussed later in "Challenges with Developing Specialty Crop Policies."

SPECIALTY CROP INSURED COVERAGE AND PARTICIPATION

Definition of *Specialty Crops*

In statute, *specialty crops* refers to "fruits and vegetables, tree nuts, dried fruits, and horticulture and nursery crops (including floriculture)" [7 U.S.C. §1621 note].[12] This definition covers more than 300 individual

[11] Some growers have expressed a preference for not developing insurance products for a number of crops. See RMA, *The Risk Management Safety Net: Portfolio Analysis-Market Penetration and Potential*, Attachment 2, August 2013.

[12] Prior to 2004, "specialty crop" were defined as "any agricultural crop, except wheat, feed grains, oilseeds, cotton, rice, peanuts, and tobacco" (Section 7(d) of the Crop Year 2001 Agricultural Economic Assistance Act (P.L. 107-111)).

agricultural commodities.[13] It includes fresh and processed fruits and vegetables, tree nuts,[14] a range of nursery plants (trees, shrubs, and flowering plants), culinary herbs and spices, coffee and tea, and also honey and maple syrup, according to guidelines established by USDA. The statutory definition of *specialty crops* ties to program eligibility and funding allocations for a number of USDA programs providing marketing and research assistance to eligible producer groups.[15]

Unlike many traditional commodity crops, specialty crops are generally not eligible for USDA farm revenue support programs—that is, programs such as the Agricultural Risk Loss Coverage (ARC), Price Loss Coverage (PLC), and Marketing Assistance Loans (MAL) programs.[16] The federal government has historically supported specialty crops indirectly through research and marketing grants.[17] The federal government provides direct assistance to individual specialty crop farmers via federal crop insurance and supplemental disaster assistance.

Overview of Premiums, Product Introductions, and Participation

Legislative changes coupled with USDA administrative efforts have contributed to broader federal crop insurance coverage of fruits, vegetables, tree nuts, and nursery crops over the past few decades. Total

---

[13] USDA estimates that this covers 46 fruits and tree nuts, 49 vegetables, 109 culinary and medicinal herbs, and 117 floriculture and nursery crops (FCIC, *Report to Congress: Specialty Crop Report*, April 2015).

[14] Peanuts do not grow on trees and are not tree nuts and therefore do not fall under the definition of *specialty crops*.

[15] For more background information, see CRS Report R44719, *Defining "Specialty Crops": A Fact Sheet*.

[16] One exception is pulse crops. Defined as dried peas, lentils, and chickpeas in the 2014 farm bill (Agricultural Act of 2014, P.L. 113-79, §§1111(6) and (17)), certain pulse crops are included in the definition of *covered commodities* eligible for ARC, PLC and MAL. These crops are also considered specialty crops under USDA's definition. See USDA, "USDA Definition of Specialty Crop," https://www.ams.usda.gov/sites/default/files/media/USDASpecialtyCropDefinition.pdf.

[17] For more information on federal support for specialty crops, see CRS Report R42771, *Fruits, Vegetables, and Other Specialty Crops: Selected Farm Bill and Federal Programs*.

liability, or the estimated value of the insured portion of the crop, is a useful measure of program growth. Specialty crop insured liabilities rose from nearly $1 billion in 1989 to nearly $18.5 billion in 2017 (Figure 1), reflecting increased production and participation.[18] In contrast, total liabilities for all federally insured crops and livestock was $106.7 billion in 2017.[19] At $18.5 billion, specialty crops represented about 17% of the federal crop insurance portfolio by liability in 2017 (Figure 1).

Table 1 provides summary statistics of federal crop insurance coverage for specialty crops for crop years 2015 through 2017. It provides total premium, premium subsidies, producer-paid premium, liabilities, indemnities (claim payments), and the number of policies earning premium.[20] Premium subsidies across all commodities (not shown in the table) totaled $6.4 billion (63% of total premium), whereas premium subsidies for specialty crops totaled $701 million (64% of total specialty crop premium) in crop year 2017.[21]

Table 1. Summary of Federal Crop Insurance Coverage, Specialty Crops (2015-2017)

	2015	2016	2017
Total Premium ($ millions)	$961	$1,[illegible]04	$1,095
Premium Subsidies ($ millions)	$616	$699	$701
Producer-Paid Premium ($ millions)	$345	$405	$394
Liabilities ($ millions)	$17,620	$19,642	$18,474
Indemnities ($ millions)	$903	$802	$831
Policies Earning Premium	60,429	60,[illegible]90	59,186

Source: CRS, using RMA Summary of Business data.

Notes: Dollars are not adjusted for inflation. Selected crops based on the statutory definition for *specialty crops* (7 U.S.C. §1621 note).

---

[18] Dollar figures are not adjusted for inflation. Liability declined some years mostly because of reduced nursery insurance following the 2008 recession that adversely affected the housing market and demand for nursery products. See K. Collins, "Crop Insurance and Specialty Crops," *Crop Insurance Today*, August 2012, pp. 35-40.

[19] CRS, using RMA Summary of Business data.

[20] Policies earning premium are policies that went into effect and for which premium was due, as opposed to policies that producers signed up for but did not ultimately purchase, or policies that were cancelled, or did not go into effect for other reasons.

[21] Ibid. Whole Farm Revenue Protection policies are counted as specialty crop policies even though they may insure nonspecialty crop and specialty crops.

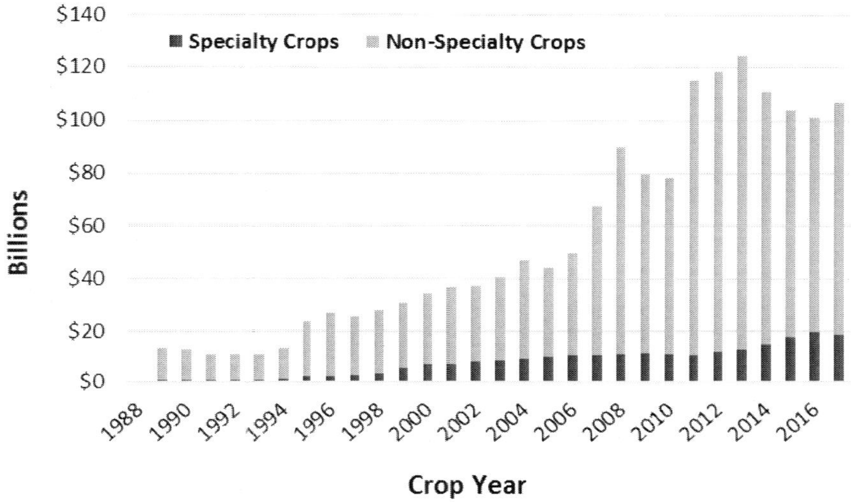

Source: CRS from USDA Risk Management Agency, Summary of Business data (accessed October 2018). Data include WFRP policies, which insure a producer's entire farm revenue and thus may include specialty crops and nonspecialty crops. Dollars are not adjusted for inflation.

Notes: Crops selected based on the statutory definition for *specialty crops* (7 U.S.C. §1621 note).

Figure 1. Total Liability for Specialty Crops Compared to All Crops, 1988-2017.

Producer-paid premiums are the difference between total premium and premium subsidies. Looking back to 1999, insurance policies then covered 52 specialty crops with planned testing on another nine, according to the Government Accountability Office (GAO).[22] These 61 crops reportedly accounted for a majority of the overall value of specialty crops at that time, but coverage on about 300 additional crops remained unavailable. Coverage expansion for specialty crops continued over the subsequent decade. By 2011, insurance was available for more than 80 specialty crops in counties considered to be major growing areas. Among the crops included were avocados, blueberries, grapes, citrus, onions, pumpkins, and tomatoes.

---

[22] GAO, *Crop Insurance: USDA's Progress in Expanding Insurance for Specialty Crops*, GAO/RCED-99-67, 1999.

The cumulative new crop insurance product introductions for specialty crops increased from 10 in 2000 to 52 in 2012 (Figure 2).[23] Figure 3 shows the number of specialty crop policies earning premium under the federal crop insurance program from crop years 2000 to 2017. While the number of specialty crop policies shows a general upward trend from 2000 to 2017, the total liability (amount insured) for specialty crops shows a more consistent upward trend during the same time period (Figure 4). A larger increase in liability than in the number of policies could be attributed to several factors, among them more high-value crop policies, increases in commodity prices, farm consolidation, and inflation.

Specialty Crops: Covered and Uncovered Crops

About 9 million acres of specialty crops, approximately 800,000 bee colonies, about 100,000 tons of raisins, and roughly 60,000 fruit and coffee trees were covered by federal crop insurance during crop year 2017. As of 2015, federal crop insurance policies were offered for 38 specialty crop categories (Table 2), which include roughly 80 types of fruits, vegetables, tree nuts, and nursery crops.[24] In 1999, insurance policies covered about 50 types of specialty crops.[25]

---

[23] More updated information is not readily available because of the difficulty of aggregating information on new product development. The RMA's online Summary of Business and Actuarial Information Browser do not indicate whether an insurance product is new. Absent an RMA report or data from private submitters, one could compare the list of all products from year to year to identify new additions and calculate the number of new products in a year.

[24] RMA data and also FCIC, *Report to Congress: Specialty Crop Report*, April 2015; and RMA, "2014 Crop Policies and Pilots." Actual estimates may vary depending on how different crop varieties are counted. A list and maps showing availability by county are available at http://www.rma.usda.gov/data/cropprograms.html.

[25] GAO, *Crop Insurance: USDA's Progress in Expanding Insurance for Specialty Crops*, GAO/RCED-99-67, 1999.

Table 2. Specialty Crop Insurance: Participation Rate and Liability, Crop Year 2015

Selected Crop	FCIC Insured Acres	Percent Eligible Acres Insured	Liability ($)
Total	8,134.316	76%	14,770,562,149
Almonds	737,312	80%	2,908,927,153
Apples	239,144	74%	1,167,133,550
Apricots	6,373	62%	17,266,826
Avocados	37,022	62%	108,596,042
Beans, fresh	4,531	6%	2,653,175
Blueberries	68,830	61%	194,126,898
Cabbage	13,500	23%	28,094,407
Cherries	88,638	68%	471,588,986
Chile peppers	1,921	10%	903,587
Citrus fruit	662,812	87%	2,670,064,107
Cranberries	32,561	80%	86,549,041
Cucumbers	30,721	25%	20,517,400
Dry beans	1,383,209	78%	537,961,368
Dry peas	1,717,930	94%	312,953,164
Figs	4,103	64%	6,518,107
Grapes/raisins	584,881	64%	1,698,101,452
Green peas	138,512	76%	62,522,493
Macadamia nuts	11,948	75%	73,077,335
Mint	19,788	22%	17,964,881
Nectarines	15,279	72%	49,915,624
Olives	26,250	73%	37,295,740
Onions	102,890	71%	200,708,027
Peaches	68,033	68%	196,749,091
Pears	33,224	72%	134,668,431
Pecans	156,867	40%	234,365,825
Peppers	8,512	20%	22,230,490
Pistachios	105,268	45%	431,109,208
Plums/prunes	57,683	86%	144,540,169
Potatoes	843,023	79%	1,131,469,456
Processing beans	94,222	57%	49,683,851
Pumpkins	8,001	19%	5,149,721
Strawberries	1,378	2%	22,033,781
Sweet corn	251,904	49%	131,363,865
Sweet potatoes	6,738	5%	4,105,253
Table grapes	78,091	63%	294,140,448
Tomatoes	335.081	82%	823,649,708
Tropical fruit	4,569	42%	35,637,868
Walnuts	153.567	53%	436,225,621

Source: CRS, using RMA Summary of Business data and NASS planted acres. Crop year 2015 represents the most complete set of acreage data for all the crop categories. Not included in total liabilities are nursery products, bee colonies, fruit/coffee trees, and tons of raisins—all of which measured in units other than acres.

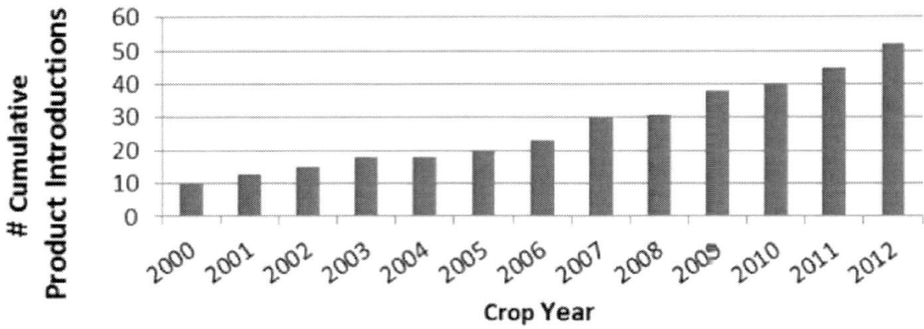

Source: K. Collins, National Crop Insurance Services, "Crop Insurance for Specialty Crops," presentation to the Specialty Crop Caucus, House of Representatives, April 12, 2012. Figure shows cumulative number of new products each year (2000-2012).

Notes: The RMA website's Summary of Business and Actuarial Information Browser does not have fields for the year a product is first introduced. Absent a custom RMA report or data from private submitters, one could compare the list of all products from year to year to identify new additions and calculate the number of new products in a given year.

Figure 2. Cumulative New Insurance Product Introductions, Specialty Crops.

Crops covered by individual federal crop insurance plans include almonds, apples, avocados, bananas, blueberries, cabbage, chili peppers, various citrus fruits, coffee, cranberries, cucumbers, dry beans, dry peas, figs, fresh market beans, fresh market sweet corn, fresh market tomatoes, grapes, green peas, macadamia nuts, mint, mustard, nursery, olives, onions, papaya, pears, pecans, peppers, pistachios, popcorn, potatoes, processing beans, pumpkins, raisins, policies on stone fruit (cherries, fresh apricots, fresh freestone peaches, fresh nectarines, peaches, plums, processing apricots, processing cling peaches, processing freestone peaches, prunes), strawberries, sweet corn, sweet potatoes, table grapes, tomatoes, other types of fruit and nut trees, and walnuts.[26]

[26] Processing crops are those not intended for sale in the high-value fresh market but that undergo further value-added.

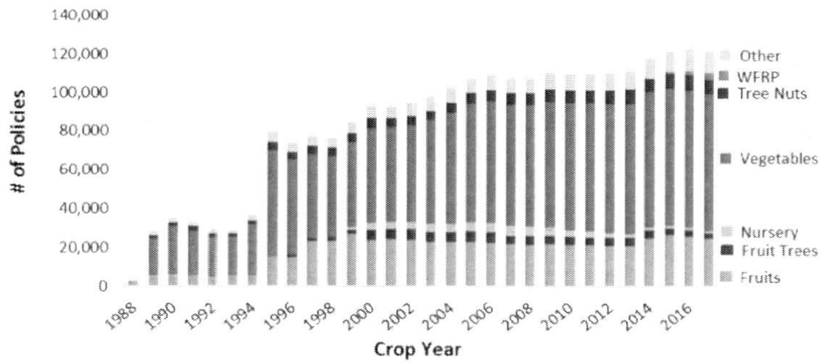

Source: CRS, using RMA Summary of Business data accessed during October 2018.

Notes: "Other" consists of camelina, clary sage, coffee, coffee trees, mint, mustard, olives, and popcorn. Selected crops based on the statutory definition of *specialty crops* (7 U.S.C. §1621 note). Policies earning premium are policies that went into effect. WFRP refers to Whole Farm Revenue Protection policies, which insure a producer's entire farm revenue and thus may include specialty crops and nonspecialty crops. RMA first offered WFRP policies in crop year 2015. Prior to that, RMA offered a similar policy known as Adjusted Gross Revenue (AGR) and AGR Lite. CRS could not locate data on AGR and AGR Lite policies on the RMA Summary of Business tool.

Figure 3. Number of Specialty Crop Policies Earning Premium.

Bee colonies are also covered. Although the availability of federal crop insurance has been expanding, it is not available for all specialty crops. Crops without insurance for which the USDA's National Agricultural Statistics Service (NASS) reports planted acreage include artichokes, asparagus, blackberries, boysenberries, broccoli, cantaloupes, carrots (fresh and for processing), cauliflower, celery, dates, garlic, guavas, hazelnuts, honeydews, kiwi fruit, lettuce, spinach, squash, tart cherries, and watermelons. In addition, specialty crops for which NASS does not report planted acreage that do not have crop-specific policies include cashews, chives, dates, eggplants, garlic, hazelnuts, leeks, lettuce, melons, most leafy greens, most herbs and spices, some tropical plants, and most root crops.

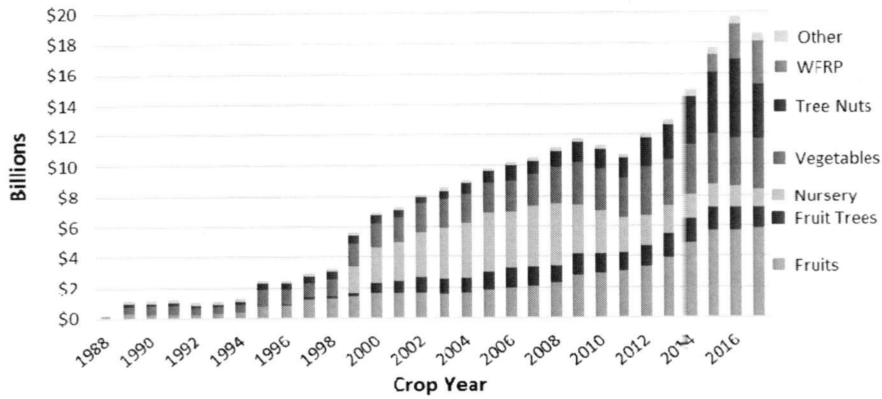

Source: CRS, using RMA Summary of Business data during October 2018.

Notes: Dollars are not adjusted for inflation. "Other" consists of camelina, clary sage, coffee, coffee trees, mint, mustard, olives, and popcorn. Dollars are not adjusted for inflation. WFRP refers to Whole Farm Revenue Protection policies, which insure a producer's entire farm revenue and thus may include specialty crops and nonspecialty crops. RMA first offered WFRP policies in crop year 2015. Prior to that, RMA offered a similar policy known as Adjusted Gross Revenue (AGR) and AGR Lite. CRS could not locate data on AGR and AGR Lite policies on the RMA Summary of Business tool.

Figure 4. Liabilities for Specialty Crops, by Type of Policy.

Market Penetration

RMA reports that nearly 300 million acres were covered by federal crop insurance policies in crop year 2015.[27] Of those acres, about 8.1 million acres (3%) were specialty crops (Table 2). For the 38 specialty crop categories for which federal crop insurance policies are measured by acreage (as opposed to bee colonies, tons, trees, or plant inventory value), the average participation rate in federal crop insurance was about 76% of eligible acres in crop year 2015.[28]

---

[27] CRS, from RMA data (https://prodwebnlb.rma.usda.gov/apps/SummaryofBusiness/ReportGenerator). Crop year 2015 represents the most complete set of acreage data for all the crop categories.

[28] Nationally, total specialty crop acreage is about 14 million acres. See USDA, *2012 Census of Agriculture*, Specialty Crops, AC-12-S-8, February 2015.

Figure 5 and Figure 6 show these data for selected fruit, tree nut, and vegetable categories. Market penetration often varies widely by crop. For fruits and tree nuts, the share of federally insured acres ranges from less than 1% (strawberries) to more than 80% (citrus and plums/prunes) (Figure 5). For vegetables and melons, insured acres range from 5% of total acres (fresh beans and sweet potatoes) to more than 80% (dry peas, tomatoes) (Figure 6). Insured acreage as a share of crop acreage is relatively high (about 70% of total specialty crop area) in major specialty crop states, including California, Florida, and Washington.[29]

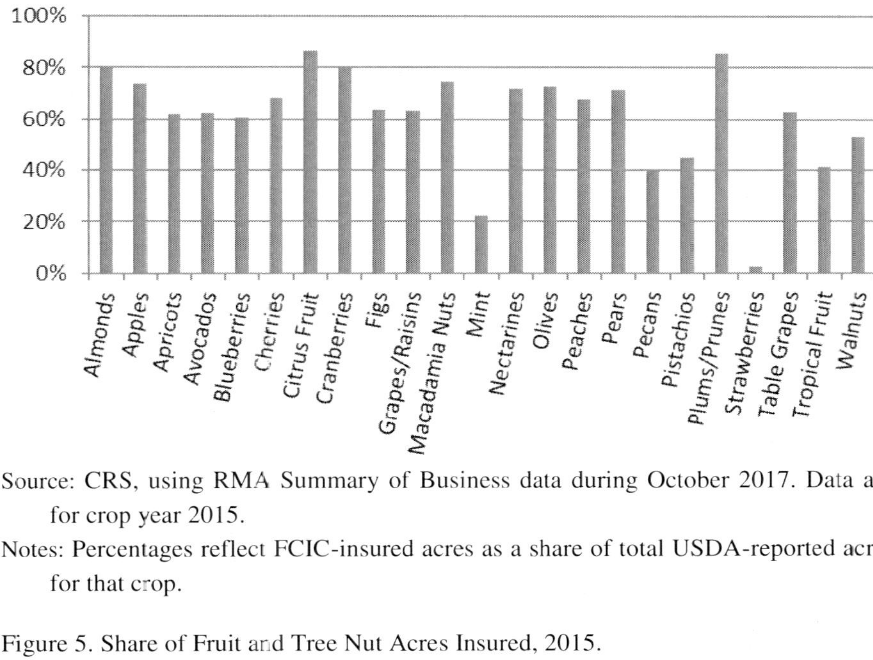

Source: CRS, using RMA Summary of Business data during October 2017. Data are for crop year 2015.

Notes: Percentages reflect FCIC-insured acres as a share of total USDA-reported acres for that crop.

Figure 5. Share of Fruit and Tree Nut Acres Insured, 2015.

[29] FCIC, *Report to Congress: Specialty Crop Report*, April 2015, https://legacy.rma.usda.gov/pubs/2015/ specialtycrop.pdf.

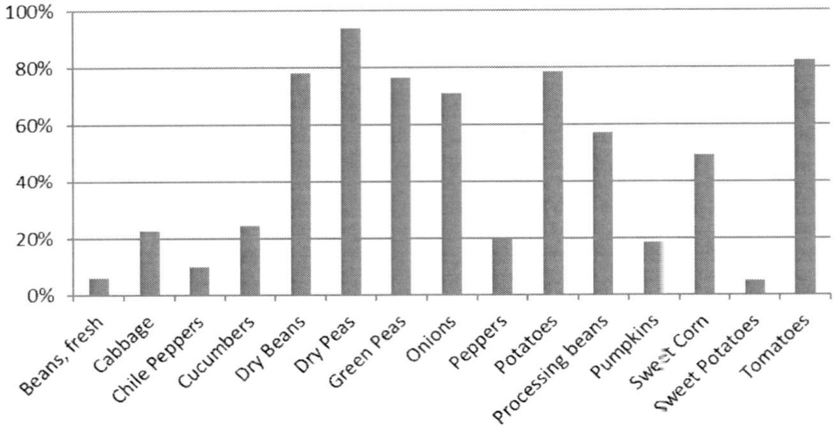

Source: CRS, using RMA Summary of Business data during October 2017. Data are for crop year 2015.

Notes: Percentages reflect FCIC-insured acres as a share of total USDA-reported acres for that crop.

Figure 6. Share of Vegetable and Melon Acres Insured, 2015.

Acreage participation for pulse crops (e.g., dry peas, dry beans) is high in Minnesota, Montana, North Dakota, and South Dakota. The 2015 FCIC report to Congress on specialty crops contains detailed acreage data by crop and state, along with maps showing crop insurance participation. Figure 7 shows liabilities for specialty crops with the highest liabilities for crop year 2017. Figure 8 shows a breakdown of crop insurance premium subsidies and producer-paid premium by specialty crop in crop year 2017. That year, the average premium subsidy amount across all specialty crop policies and WFRP was 64%, about 1% higher than the average premium subsidy rate for the entire crop insurance portfolio.

Whole Farm Revenue Protection (WFRP) Insurance

RMA provides a WFRP policy option to growers insuring 50% to 85% of revenue for all the commodities on a farm under one insurance policy. WFRP benefits farms with specialty or organic commodities that lack individual policy coverage, as well as those farms marketing to local, regional, farm-identity preserved, or direct markets. RMA estimates that 2,700 such policies received premium subsidies in 2017, up from 1,100 policies in 2015 (Figure 9). WFRP premium subsidies totaled about $102 million in 2017, and producer-paid premium amounted to about $42 million, while total liability was estimated at $2.8 billion, up from $1.1 billion in 2015.

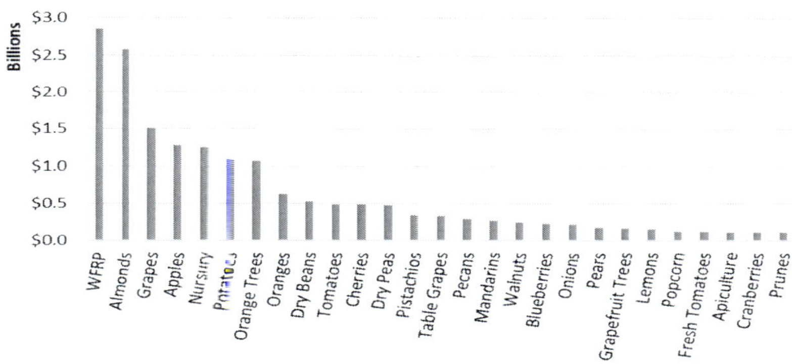

Source: CRS, using RMA Summary of Business data during October 2018. Data are for crop year 2017. The category "Tomatoes" is comprised of canning tomatoes and other tomatoes not marketed as fresh produce.

Notes: Dollars are not adjusted for inflation. WFRP refers to Whole Farm Revenue Protection policies. WFRP policies insure a producer's entire farm revenue and thus may include specialty crops and nonspecialty crops. RMA first offered WFRP policies in crop year 2015. Prior to that, RMA offered a similar policy known as Adjusted Gross Revenue (AGR) and AGR Lite. CRS could not locate data on AGR and AGR Lite policies on the RMA Summary of Business tool. The figure includes all specialty crops (and WFRP policies) that had over $100 million in liability. WFRP and the specific crops in the figure accounted for 94% of all federal crop insurance liabilities for WFRP and specialty crops.

Figure 7. Specialty Crops with the Highest Total Liabilities, 2017.

The 2014 farm bill (P.L. 113-79) required that RMA provide a WFRP policy option to agricultural producers and authorized higher premium subsidy levels for whole-farm policies than for other policy types. The WFRP pilot policy was first offered in 2015.

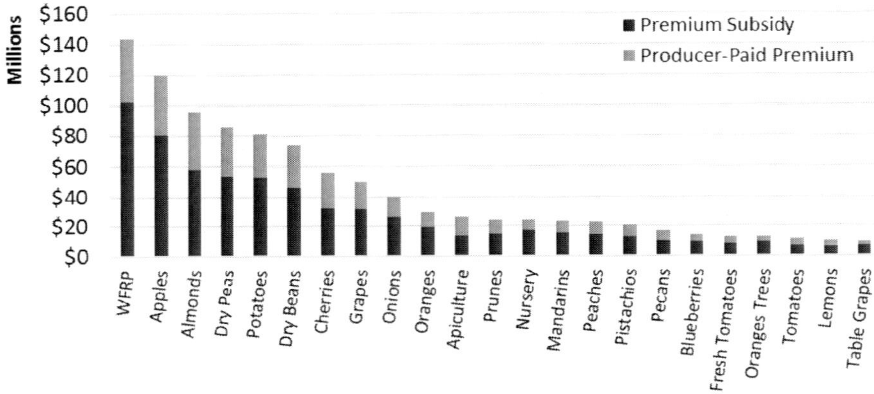

Source: CRS, using RMA Summary of Business data during October 2018. Data are for crop year 2017. The category "Tomatoes" is comprised of canning tomatoes and other tomatoes not marketed as fresh produce.

Notes: Selected fruit and vegetable product categories, including herbs and tropical fruits (excluding apiaries).

Figure 8. Specialty Crop Premium Subsidies and Producer-Paid Premium, 2017.

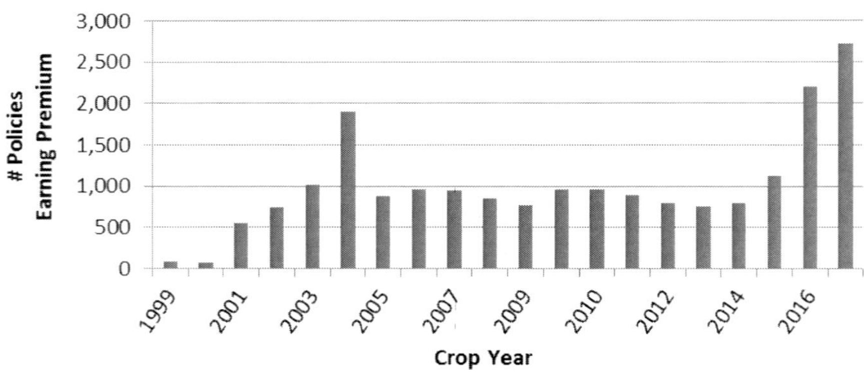

Source: CRS, using Risk Management Agency Summary of Business data accessed during October 2017.

Notes: Data are not available to distinguish between policies for specialty crop growers from all U.S. agricultural producers. Available data reflect: Adjusted Gross Revenue (AGR) policies (1999-2002), AGR and AGR Lite policies (2003-2014), and WFRP policies (2015-2017).

Figure 9. Whole Farm Type Policies.

Prior to that, USDA offered similar policies under other names (in particular, Adjusted Gross Revenue or "AGR" policies from 1999 to 2002, and AGR and AGR Lite policies from 2003 to 2014). WFRP insures 50% to 85% of revenue for all commodities on a farm under one insurance policy, including (but not limited to) farms with specialty or organic commodities or those marketing to local, regional, farm-identity preserved, specialty, or direct markets. WFRP is available in all counties nationwide—either as a stand-alone policy or in combination with other policies—to farms with up to $10 million in insured revenue at the 85% coverage level and up to $17 million in insured revenue at the 50% coverage level.[30] WFRP does not offer CAT-level coverage (high deductible, 100% premium subsidy) and cannot be combined with CAT-level policies. WFRP premium subsidies range from 55% to 80%, and coverage levels range from 50% to 85%. In crop year 2017, the average premium subsidy for WFRP was 70%, whereas the average premium subsidy across the entire crop insurance portfolio was 63%. That year WFRP policies accounted for 3% of all federal crop insurance liabilities.

WFRP requires producers to provide five consecutive years of Schedule F from their federal tax forms on which farm income and expenses are reported to determine their historical revenue guarantee. Some speculate that one of the reasons WFRP participation is not higher is that producers are reluctant to provide tax data.[31]

SELECT POLICY TYPES RELEVANT TO SPECIALTY CROPS

Federal crop insurance policies for specialty crops (and all other crops) are generally either yield-based or revenue-based. For most yield-based policies, a producer can receive an indemnity if there is a yield loss relative

[30] For more information, see RMA, "Whole-Farm Revenue Protection," September 2018; and RMA, "Frequently Asked Questions: Whole-Farm Revenue Protection (WFRP) Plan," September 2018.

[31] See, for example, A. Johnson et al., "New Option for Farm Risk Management: Whole Farm Revenue Protection Usage in Nebraska," p. 10, September 2017, https://www.cfra.org/sites/www.cfra.org/files/publications/ WFRP%20Report.pdf.

to the farmer's "normal" (historical) yield. Revenue-based policies protect against crop revenue loss resulting from declines in yield, price, or both.[32]

Yield-Based Insurance Policies

There are two types of yield-based insurance policies for specialty crops: (1) Actual Production History (APH) plans and (2) dollar plans. APH policies account for about 70% of specialty crops policies, with the exception of nursery crops, which tend to be mostly covered through dollar plans.[33] Text Box 1 shows examples of an APH policy for citrus fruit with a 50% coverage level and a dollar plan for nursery products with a 65% coverage level.

Actual Production History (APH) Plans

APH policies insure producers against yield losses due to natural causes such as drought, excess precipitation, hail, frost, freeze, fire (if due to natural causes), and insects and disease. An indemnity is not paid if crop loss is caused by insufficient or improper applications of pest or disease control measures.[34] The producer selects the amount of average yield to insure, ranging from 50% to 75% (in some areas up to 85%). The producer also selects the percent of the predicted price to insure, ranging between 55% and 100% of the crop price established annually by RMA. If the harvested crop plus any appraised production is less than the yield insured, the producer is paid an indemnity based on the difference. Indemnities are calculated by multiplying this difference by the insured percentage of the price selected when crop insurance was purchased and by the insured share (coverage level).

---

[32] For additional background, see USDA, "Insurance Plans," https://www.rma.usda.gov/policies/.

[33] J. Lee, "Crop Insurance for 'Specialty' Crops," presented at the Farm Bill and Western Agriculture workshop, 2013.

[34] For some crops, such as California and Arizona navel oranges and California avocadoes, RMA has a Quarantine Endorsement Pilot Program that, for an additional premium, provides insurance protection when a quarantine affects all or part of the insured crop. In some areas, an endorsement is available to protect against loss due to quarantine established to control a specific pest that requires destruction of insured crops.

When purchasing an APH policy, a producer is assigned a "normal" crop yield based on the producer's actual production history and a price for the commodity based on estimated market conditions. The producer can then select a percentage of his normal yield to be insured and a percentage of the price he or she wishes to receive when crop losses exceed the selected loss threshold.[35]

Dollar Plans

Dollar plans provide protection against declining value due to damage that causes a yield shortfall. Unlike APH policies, a dollar policy guarantees a dollar amount of coverage and not a level of production, with the amount of insurance based on the cost of growing a crop in a specific area. A loss occurs when the annual crop value is less than the amount of insurance. The maximum dollar amount of insurance is stated on the actuarial document.[36] The insured may select a percentage of the maximum dollar amount equal to CAT or higher coverage levels.

The design and implementation of dollar plan policies have been criticized for not insuring actual losses and for covering fraudulent claims. A 1997 USDA Office of the Inspector General audit report on fresh market tomato dollar plans outlined several specific fraud, waste, and abuse concerns.[37] More recently, in December 2017 an RMA-commissioned report on options for improving or replacing dollar plans concluded that the dollar plan as studied (limited area and three crops) "is not sustainable."[38]

---

[35] A detailed explanation of crop insurance, with examples, is available in RMA, *Crop Insurance Option for Vegetable Growers*, October 2008.

[36] RMA, "Insurance Plans," https://www.rma.usda.gov/policies/.

[37] USDA, Office of the Inspector General, *Crop Insurance on Fresh Market Tomatoes, Crop Year 1996—Florida*, Audit Report No. 05099-1-AT, September 1997, https://www.usda.gov/oig/webdocs/05099-1-AT.pdf.

[38] Watts and Associates, "Dollar Insurance Recommended Modifications Report Final," December 14, 2017, p. 3, https://legacy.rma.usda.gov/pubs/2018/dollarplanmodification.pdf.

Text Box 1: Actual Production History (APH) and Dollar Plan

APH Policy for Citrus Fruit: 50% Coverage Level

A claim can be filed whenever production falls short of the guarantee selected by the insured. The indemnity is determined by multiplying the production shortfall by the pre-selected price. The following example uses an APH yield of 600 cartons per acre, 50% coverage level, and a selected price of $5.55 per carton.

600 cartons per acre average yield (APH yield)
x 0.50 selected coverage level percentage
300 cartons per acre guarantee
- 200 cartons per acre actually produced
100 cartons per acre loss
x $5.55 price election
$555 gross indemnity per acre

Source: USDA Fact sheet, "Citrus–Actual Production History," June 2012.

Dollar Plan for Nursery: 65% Coverage Level

The dollar plan provides protection against a decline in value due to damage that causes a yield shortfall. The following example uses a plant inventory value of $100,000 and a 65% coverage level.

$100,000 plant inventory value
x 0.65 selected coverage level percentage
$65,000 unit amount of insurance
In the event of a loss:
$100,000 field market value before loss
-$50,000 field market value after loss
$50,000 value of loss
-$35,000 deductible (1.0 - coverage level) X inventory = (1 - 0.65) X $100,000 = $35,000
$15,000 indemnity

Source: USDA, "Nursery Commodity Insurance," February 2018.

Revenue-Based Insurance

Revenue insurance is widely available for major program crops (e.g., wheat, corn, soybeans) and protects growers against losses from low yields, low prices, low quality, or any combination of these events. For specialty crops, designing revenue-based insurance products is challenging. These crops often do not have centralized price discovery mechanism such as a futures exchange for developing price projections prior to planting. They also often lack data on actual harvest-time prices.

Text Box 2: Actual Revenue History (ARH) for California Navel Oranges

ARH protects against losses from low yields, low prices, low quality, or any combination of these events.

Coverage is based on the farmer's own net revenue history and the farmer's choice of coverage level and payment factor (which affects the premium cost and potential indemnity). The crop's revenue is determined after harvest at the point of first delivery.

Insurance Terms:
$3,200 per acre approved revenue
x 0.75 selected coverage level percentage
x 1.0 selected payment factor (ranges from 0.67 to 1.0; affects the premium cost to producer)
$2,400 per acre amount of insurance

Actual Revenue:
400 cartons
x 5.00 per carton net price
$2,000 per acre revenue to count

Indemnity:
$2,400 per acre amount of insurance
-$2,000 per acre revenue to count (actual revenue)
$400 difference
x 1.0 payment factor
$400 indemnity per acre

Source: USDA Fact Sheet, "Actual Revenue History-Navels," March 2012.

To address these types of data challenges, actual revenue history (ARH) insurance plans have been implemented on a pilot basis for certain specialty crops such as navel oranges and cherries. Rather than insuring historical yields, these pilot policies insure historical revenues using historical prices. This approach assumes that historical prices provide a reasonable estimate of expected future prices. This assumption is deemed viable for perishable crops, such as most fruit and vegetables, but is considered less tenable with storable crops where stock carryover from the previous year can affect current market-year prices.[39] ARH insurance plans have parallels to the APH insurance plans, with the primary difference being that instead of insuring historical yields, the plan insures historical revenues.[40] Text Box 2 shows an example of an ARH policy for California navel oranges.

Designed as a catch-all for a variety of crops that may not have individual revenue insurance plans, WFRP policies insure 50%-85% of revenue for all commodities on a farm under one policy. For more information on WFRP, see "Whole Farm Revenue Protection (WFRP) Insurance."

Index-Based Policies

RMA offers a few index-based policies, which trigger claim payments based on a predetermined index that is entirely independent of the individual farm operation (e.g., rainfall level). Indemnities are automatically triggered whenever the index falls below a producer-selected coverage level instead of requiring insureds to file claims.[41] One of those policies covers a specialty crop—the Apiculture Pilot Insurance Program (API), which covers honey production. Text Box 3 provides detailed information on producers' choices under API.

---

[39] FCIC, *Report to Congress: Specialty Crop Report*, November 2010.

[40] RMA, "Insurance Plans," https://www.rma.usda.gov/policies/.

[41] For more information on index insurance, see World Bank Group, International Finance Corporation, "Index Insurance—Frequently Asked Questions," https://www.ifc.org/wps/wcm/connect/industry_ext_content/ifc_external_corporate_site/industries/financial+markets/retail+finance/insurance/index+insurance+- +frequently+asked+questions.

Text Box 3: Apiculture Rainfall Index Policy

API provides a safety net for beekeepers' primary income source—honey, pollen collection, wax, and breeding stock—based on a rainfall index. Apiculture systems consist of different types of plants or crops and often contain mixtures of different species, each with different growth habits and seasons, precipitation requirements, and other climate conditions necessary to maintain plant growth over extended periods of time. API was designed to provide maximum flexibility to cover these diverse situations.

The Rainfall Index uses National Oceanic and Atmospheric Administration Climate Prediction Center data, which uses a grid system to evaluate the geographic location of specific index values. Each grid is 0.25 degrees in latitude by 0.25 degrees in longitude, which translates to approximately 17 by 17 miles at the equator. Colonies will be assigned to one or more grids based on the location to be insured.

Availability

API is available in the 48 contiguous states with the exception of grids that cross international borders.

Coverage and Claims

Coverage is based on producers' selection of coverage level, index intervals, and productivity factors. The index interval represents a two-month period, and the period producers select should be the one when precipitation is most important to their operations.

Coverage levels vary from 70% to 90% of the maximum dollar amount of protection offered (total liability) for a producer's colonies. In other words, the policy requires a deductible ranging from 10% to 30%.

Producers select a productivity factor to match the amount of protection to the value of the production that best represents their operations and the productive capacity of their colonies. They do not have to insure all of their colonies. However, they cannot insure more than the total number of colonies they own.

By selecting a productivity factor, producers can establish a value between 60% and 150% of the county base value, which is based on honey production and uses a five-year rolling average of NASS data. The yield data are based on the NASS state average, and the price is the national average honey price for a given year.

Producers' insurance payments are determined by using Climate Prediction Center data for the grid(s) and index interval(s) they have chosen to insure. When the final grid index falls below their "trigger grid index," they may receive an indemnity. This insurance coverage is for a single peril: lack of precipitation. Coverage is based on the experience of the entire grid. It is not based on an individual farm or ranch or specific weather stations in the general area.

Tools

Producers are asked to make several choices when insuring production from their colonies, including coverage level, index intervals, productivity factor, and number of colonies. Producers may work with their crop insurance agents to view the Grid ID Locator map and index grids for their area.

Source: RMA Fact Sheet, "Apiculture Pilot Insurance Program—Honey, Pollen Collection, Beeswax, Breeding Stock," https://legacy.rma.usda.gov/pubs/rme/apiculture.pdf.

DEVELOPMENT OF NEW POLICIES AND SECTION 508(H)

The Federal Crop Insurance Act provides two methods for developing new crop insurance programs, including (1) internal products developed by RMA or under contract and (2) external products submitted through procedures specified in Section 508(h) of the Federal Crop Insurance Act (7 U.S.C. §1508(h)).

RMA-Developed Products

Section 522 of the Federal Crop Insurance Act (7 U.S.C. §1522) grants RMA authority to develop new crop insurance policies. This authority was partially removed in 2000 but later reinstated by the 2014 farm bill (P.L. 113-79). Before the enactment of Agricultural Risk Protection Act of 2000 (ARPA, P.L. 106-224), products were typically developed internally.[42] ARPA added paragraph (e)(4) to Section 522, which stated, "on and after October 1, 2000, the Corporation shall not conduct research and development for any new policy for an agricultural commodity offered under this subchapter." The 2014 farm bill repealed paragraph (e)(4).

[42] CRS communication with RMA.

Private Sector Developed Products

ARPA expanded the role of the private sector, allowing private entities to participate in conducting research and development of new insurance products and features. With the expansion of the contracting and partnering authority, Congress authorized RMA to enter into contracts or to create partnerships for research and development of new and innovative insurance products. Private entities could also submit unsolicited proposals for insurance products to the FCIC Board of Directors for approval. In considering such proposals, the board is to evaluate whether the products (1) are in the best interests of producers, (2) follow sound insurance principles, and (3) are actuarially appropriate.[43]

Section 508(h) governs new crop insurance policy development, including how it is contracted out and funded, how policy ratings are undertaken, and how a policy may start as a pilot and may (or may not) evolve to a full-fledged insurance policy. FCIC is authorized to reimburse private entities for research, development, and maintenance costs if they develop insurance programs that are approved by the FCIC board.[44]

Private sector individuals may submit to the FCIC board (1) crop insurance policies, (2) provisions of policies, or (3) rates of premium. These submissions are commonly referred to as 508(h) submissions. If a private individual prefers, a concept proposal can be submitted to the board prior to fully developing a 508(h) submission. The board may approve an advance payment for the concept proposal for up to 75% of expected research and development expenses to aid in the development of the 508(h) submission.[45]

---

[43] RMA, "Private Sector Developed Plans of Insurance," https://legacy.rma.usda.gov/fcic/privatesector.html.

[44] Section 522(c) provides contracting authority, stating the FCIC "may enter into contracts to carry out research and development."

[45] The 2008 farm bill first authorized the FCIC board to approve advance partial payments for the development of new products. Section 522(b)(2)(A) states that the board may approve the applicant's request for advanced payment of a portion of the research and development costs prior to submission and approval of the policy by the board (under Section 508(h). FCIC procedures specify: "At any time after an applicant has been approved for an initial advance payment for research and development costs of up to 50 percent, the applicant may submit a one-time request for an additional advance payment of up to 25 percent by

If approved by the board, these insurance products can receive reimbursement for research, development and operating costs, in addition to any approved premium subsidies and reinsurance. Private submitters are eligible to recoup maintenance costs for up to three years after products are offered on the market if they continue to provide support for the products.[46] After three years, the private entity has the choice of turning the product over to RMA, thereby relinquishing all ownership rights in the product, or retaining ownership of the insurance product and continuing to update it in return for a user fee as approved by the board and paid by Approved Insurance Providers who sell the product.[47]

Pilot Status of New Products

The FCIC board must approve all new products. This process can take up to a year and generally depends on the quality and thoroughness of the submission package presented to the board, as well as the responsiveness of the submitter to issues raised by the board and the reviewers, among other factors. In most cases, an independent external panel of experts reviews a proposed product and assesses its actuarial weakness and suggests product improvements. The revised product is submitted to the FCIC board for approval. Once approved, the product is typically implemented as a pilot program in a limited area to test it for effectiveness while limiting financial exposure.

Pilot programs typically operate for four years but may be extended for additional testing if needed.[48] Eventually the FCIC board either converts the pilot to a regular program or terminates it. Under law, RMA is not

submitting a written request." See RMA, "Approved Procedures for Submission of Concept Proposals Seeking Advance Payment of Research and Development Costs," FCIC-17030, June 13, 2017, p. 8, https://legacy.rma.usda.gov/fcic/2017/17030.pdf.

[46] The submitter may relinquish support of the product, thus surrendering any additional year's maintenance fees.

[47] RMA, "History of RMA," https://www.rma.usda.gov/About-RMA/History-of-RMA.

[48] Section 523(a)(1) states that the FCIC may conduct a pilot program submitted to and approved by the board under Section 508(h) or that is developed under Section 522(b) in order to "evaluate whether a proposal or new risk management tool tested by the pilot program is suitable for the marketplace and addresses the needs of producers."

allowed to conduct pilot programs if insurance against the risk to be covered is generally available in the private sector without governmental support.

RMA's 2010 *Report to Congress* describes selected pilot programs that have been developed through the internal procedure and authority, covering quarantine and policies based on actual revenue history (e.g., rather than yields).[49] It also describes a range of private sector initiatives covering pumpkins, apiculture, plantains and bananas, sugarcane, and fresh market beans.

CHALLENGES WITH DEVELOPING SPECIALTY CROP POLICIES

Even though new crop insurance product introductions for specialty crops have been increasing, USDA and the industry continue to face a number of challenges when developing and making available new insurance policies for specialty crops that are not currently covered. Most challenges stem from the basic structure of the specialty crop industry, which is often characterized by relatively small acreages, multiple crop varieties (often targeting niche markets), differences in farming practices (which contribute to greater complexity and cost), quality and price discovery issues, grower interest, nonweather risks, and other coverage limitations.[50] Factors such as these affect the potential marketability, actuarial soundness, and feasibility of an insurance policy.

A small market reduces sales incentive for companies selling insurance while contributing to higher per-unit costs for developing the product,

---

[49] RMA, *Report to Congress: Specialty Crop Report*, November 2010.

[50] Information presented here is excerpted in part from K. Collins, National Crop Insurance Services, "Crop Insurance for Specialty Crops," presentation to the Specialty Crop Caucus, House of Representatives, April 12, 2012. Other information is from FCIC, *Report to Congress: Specialty Crop Report*, April 2015, https://www.rma.usda.gov/pubs/2015/specialtycrop.pdf; J. Seufer, RMA, *Managing Nursery Production Risks*; and M. S. Paggi, "The Use of Crop insurance in Specialty Crop Agriculture," *CHOICES*, 2016, http://www.choicesmagazine.org/choices-magazine/themearticles/crop-insurance-in-the-20182019-farm-bill/the-use-of-crop-insurance-in-specialty-crop-agriculture.

training staff, modifying computer programs, and other activities. Small acreage also results in low market volume or the establishment of production contracts between producers and buyers. Crops grown and marketed in smaller quantities and/or targeting niche markets often command a price premium, resulting in often highly variable market prices, further complicating price discovery. Moreover, most specialty crops are intended primarily for sale in the higher-value fresh market versus the typically lower-value crops sold for further processing. Fresh product is highly perishable and nonstorable, unlike the field crops that are more widely covered by federal crop insurance policies.

In general, lack of reliable pricing data for crops not traded on commodity exchanges has been an ongoing challenge for USDA in the federal crop insurance program.[51] RMA's price discovery for specialty crops largely relies on Agricultural Marketing Service, NASS, and other USDA agency data and academic and industry sources.[52] In the absence of a well-developed cash market, such "thin market" conditions make it difficult to observe and forecast market prices. For example, in 2015 FCIC cancelled the Dry Bean Revenue Endorsement because USDA did not have sufficient market data from processors to establish a harvest price from which to calculate whether indemnities would be triggered under the endorsement.[53]

Setting price guarantees correctly is critical for encouraging participation, the actuarial soundness of the program, and maintaining the overall market dynamics for the crop. If the insurance is priced (rated) too high, producers who tend to have few losses might decide against purchasing insurance, leaving only high-risk farmers in the pool (known as "adverse selection"). If the insurance is priced too low, premiums may not

---

[51] Collins, "Crop Insurance and Specialty Crops," p. 40.

[52] RMA, "A Report from the Risk Management Agency, U.S. Department of Agriculture, regarding: The Department of Agriculture's Annual Report to the Committee on Agriculture of the U.S. House of Representatives and the Committee on Agriculture, Nutrition and Forestry of the U.S. Senate Regarding the Progress Made in Developing and Improving Federal Crop Insurance for Organic Crops," March 2017.

[53] For more information, see Gregory Ackerman v. U.S. Department of Agriculture, No. 17-cv-11779, 2018 WL 1858165 (E.D. Mich. April 18, 2018), a lawsuit filed by a group of dry bean farmers from three states that purchased the Dry Bean Revenue Endorsement in crop year 2015.

cover expected indemnities, potentially inflating the federal cost of the program by providing a greater premium and higher administrative and operating subsidies than was intended. Artificially low premiums might encourage additional crop production, further contributing to weak market prices, thereby adversely affecting financial returns for producers.

Another challenge for insuring specialty crops is the diversity and multitude of crop varieties and production practices. Compared with field crops, specialty crops tend to have a wider variety of farming practices that depend on the crop variety, adding complexity to the policy and its development cost. For example, a vegetable crop may need to be grown on raised beds, use plastic, or have specific crop rotations. Various marketing claims made for products (such as that they are organic or other production or sustainability claims) can contribute to product complexity. Understanding how these factors affect potential yields is required for determining what practices can be insured and for developing and establishing underwriting standards. Variation across crops and variety within crop types also complicate the loss adjustment process (i.e., assessing the effect of weather on crop production).

Finally, in some cases, USDA has reportedly not pursued policies for particular commodities because some producers have expressed concerns that offering insurance could adversely affect the market (i.e., because an insurance policy reduces producer risk, farmers may plant more acreage, which could drive down prices and total crop revenue). This has been a particular concern for producers of vegetable crops and explains in part lower levels of insured vegetable crop acreage compared with other crops.[54]

Producer interest in the availability of a new policy often starts at the local level and is channeled through RMA's regional offices. In general, for a policy to be viable, a crop must have established cultivars, defined farming practices, developed markets, and identified known perils. Significant producer interest (demand for the policy) is also critical.

---

54 Some growers have expressed a preference for no development of insurance products for a number of crops. See RMA, *The Risk Management Safety Net: Portfolio Analysis-Market Penetration and Potential*, Att. 2, August 2013.

Perhaps more importantly, insurance policies are dependent on the availability of high quality data. *High quality data*, from an insurance standpoint, refers to data being timely (so that claims can be paid quickly), relevant (so the product offers reliable protection), audited to international reinsurance standards, and available over a sufficiently long time horizon (time series).[55] Data availability and data quality often pose a challenge for crop insurance purposes.

ALTERNATIVES TO CROP INSURANCE

USDA offers several programs to help farmers recover financially from natural disasters, including droughts and floods. These programs help to provide assistance to producers of noninsured crops or crops with no current individual policy coverage, including some specialty crops. Other supplemental disaster programs further provide assistance to some specialty crop producers from tree losses and the loss of bee colonies. USDA also provides low-interest emergency loans and land rehabilitation assistance to help farmers return land to production following natural disasters.[56]

CONSIDERATIONS FOR CONGRESS

Among the issues that may arise if Congress continues to consider the role of the federal crop insurance options for specialty crop producers are:

- **Availability of crop-specific policies.** Crop-specific policies have not been developed for a number of specialty crops, including artichokes, asparagus, blackberries, boysenberries, broccoli,

[55] World Bank Group, *Agricultural Data and Insurance*, September 2015.

[56] All the programs have permanent authorization, and only the emergency loan program requires a federal disaster designation. For more information on these programs, see CRS Report RS21212, *Agricultural Disaster Assistance*, and CRS Report R42854, *Emergency Assistance for Agricultural Land Rehabilitation*.

cantaloupes, carrots (fresh and for processing), cashews, cauliflower, chives, celery, dates, eggplants, garlic, guavas, hazelnuts, honeydews, kiwi fruit, lettuce, spinach, squash, tart cherries, watermelons, most leafy greens, most herbs and spices, some tropical plants, and most root crops. Private submitters proposing to develop new policies must present evidence of marketability to FCIC and RMA.[57] Congress might consider whether there are opportunities for USDA to facilitate market research and publish market data to assist with the development of new policies and spur greater competition among private submitters.

- **Development cost benchmarks.** Section 11120 of P.L. 115-334 modifies the definition of *reasonable research and development costs* related to policies that have been approved by the FCIC board for reimbursement.[58] Costs are to be deemed reasonable if based on (1) for employees or contractors, wage rates equal to not more than two times the Bureau of Labor Statistics hourly wage rates, plus benefits; and (2) other actual documented costs incurred by the applicant. That section also limits the FCIC board's review of user fees. Given that Congress reinstated RMA's authority to develop new products in the 2014 farm bill (P.L. 113-79), Congress might inquire whether costs incurred for RMA-developed policies could provide a benchmark for the reasonableness of private sector requests for reimbursement of development costs.
- **Limited participation in WFRP policies.** WFRP is available in every county of every state. Its premium subsidies range from 55% to 80%, and coverage levels range from 50% to 85%. In crop year 2017 the average premium subsidy for WFRP was 70%, whereas the average premium subsidy across the entire crop insurance portfolio was 63%. That year WFRP policies accounted for about

---

[57] For detailed information on the requirements, see RMA, "Private Sector Developed Plans of Insurance," https://legacy.rma.usda.gov/fcic/privatesector.html.

[58] See also H.Rept. 115-1072.

3% of all federal crop insurance liabilities. Some speculate that part of the reason that WFRP participation is not higher is that producers are reluctant to provide tax return data.[59] These policies require producers to provide five consecutive years of Schedule F from their federal tax forms. Given the benefits of risk pooling achieved by insuring a whole farm's revenue and the fraud prevention benefits from requiring tax returns to set historical revenue guarantees, Congress might consider whether there are efficiencies to be gained from incentivizing greater participation in WFRP.

- **Determining a "market price" for commodities not sold on exchanges.** Lack of reliable pricing data for crops that are not traded on commodity exchanges has been an ongoing challenge for USDA in the federal crop insurance program. RMA's price discovery for specialty crops largely relies on Agricultural Marketing Service, NASS, and other USDA agency data and academic and industry sources. In the absence of a well-developed cash market, such "thin market" conditions make it difficult to observe and forecast market prices, which affects RMA's ability to set the appropriate level of price guarantees. Setting price guarantees correctly is critical to the actuarial soundness of the program and for maintaining overall market dynamics for the crop. Increased farm-level price data for commodities not sold on exchanges could also assist producers of those commodities in negotiating contracts and in their financial planning. Congress might consider whether there is a role for the federal government in supporting data collection of farm prices of commodities that are not sold on exchanges.
- **Coverage of quality losses.** Many crops are vulnerable to lower prices or to becoming unmarketable due to quality losses. Small markets for specialty crops that are not sold on commodity exchanges may be particularly price-sensitive to variations in

[59] E.g., Johnson et al., "New Option for Farm Risk Management," p. 10.

quality (e.g., herbs and spices, certain fruits and vegetables, honey). Certain federal crop insurance policies cover some quality losses, but the range of such coverage is limited. Congress might consider (1) whether the current coverage for quality losses is available for all crops that are vulnerable to quality losses, (2) whether loss adjustment procedures for quality losses accurately assess the variations in quality and the effects on marketability and prices, and (3) whether the procedures for assessing quality losses properly balance cost efficiency and fraud prevention.

- **Effect of ad hoc payments on demand for crop insurance.** One of the stated policy goals of federal crop insurance has been to reduce the agricultural sector's reliance on supplemental or "ad hoc" disaster assistance payments.[60] However, according to the Congressional Budget Office, it is difficult to assess whether this policy goal has been achieved.[61] The 115th Congress authorized ad hoc disaster assistance, and USDA has separately implemented a "trade aid" package, both of which apply to some specialty crops.[62] Given the lower participation levels for certain specialty crop insurance policies as compared to nonspecialty crops, Congress might consider whether repeated ad hoc payments in response to adverse events may have an effect on demand for federal crop insurance for crops generally and for specialty crops in particular.

---

[60] Congressional Budget Office, *Options to Reduce the Budgetary Costs of the Federal Crop Insurance Program*, December 2017, p. 1, https://www.cbo.gov/system/files/115th-congress-2017-2018/reports/53375-federalcropinsuranceprogram.pdf.

[61] Ibid.

[62] For example, the Bipartisan Budget Act of 2018 (P.L. 115-123) provided $2.36 billion to the Secretary of Agriculture to cover crop, tree, bush, and vine losses from 2017 hurricanes and wildfires. For more information, see CRS In Focus IF10829, *Agriculture Funding in the Bipartisan Budget Act of 2018*. Also, in 2018 USDA authorized up to $12 billion in financial assistance—referred to as a "trade aid" package—for certain agricultural commodities using Section 5 of the Commodity Credit Corporation Charter Act (15 U.S.C. 714c). For more information, see CRS Report R45310, *Farm Policy: USDA's Trade Aid Package*.

In: Agricultural Issues
Editor: Michael Butlin
ISBN: 978-1-53615-474-0

Chapter 9

ANIMAL AND PLANT HEALTH IMPORT PERMITS IN U.S. AGRICULTURAL TRADE*

Sahar Angadjivand

SUMMARY

The Animal and Plant Health Inspection Service (APHIS) of the U.S. Department of Agriculture (USDA) is the U.S. government authority tasked with regulating the import, transit, and release of regulated animals, animal products, veterinary biologics, plants, plant products, pests, organisms, soil, and genetically engineered organisms. APHIS provides scientific authorities in trade partner countries and U.S. importers with animal and plant health import regulations.

APHIS requires U.S. importers to obtain animal or plant health import permits, which verify that the items being imported meet U.S. import standards. Animal and plant health import permits certify that imports follow U.S. regulations, World Trade Organization (WTO) guidelines, and/or trading partner specific requirements. These import permits are a part of broader agreements between the United States and its trading partners within the WTO on established sanitary and

* This is an edited, reformatted and augmented version of Congressional Research Service Publication No. .R45457, updated January 11, 2019.

phytosanitary (SPS) measures. These measures aim to protect against diseases, pests, toxins, and other contaminants. The House and Senate Agricultural Appropriations Committees appropriate funds that allow APHIS to carry out a range of activities, including those involved in issuing import permits. From FY2014 to FY2018, discretionary appropriations for APHIS have averaged nearly $900 million. About 85% of the APHIS budget is allocated to the "Safeguarding and Emergency Preparedness/Response" mission area, which includes the administration of health import permits and other efforts to prevent imports of pests and diseases into the United States.

APHIS's authority over agricultural imports is largely provided by the Animal Health Protection Act (7 U.S.C. §§8301 *et seq.*), the Plant Protection Act (7 U.S.C. §§7701 *et seq.*), and the Public Health Security and Bioterrorism Preparedness and Response Act of 2002 (7 U.S.C. §§8401). These laws authorize APHIS to administer animal and plant health import permits and conduct agricultural import inspections. APHIS works with other federal agencies, such as the Department of Homeland Security's Customs and Border Protection (CBP), to conduct animal and plant health monitoring programs and to determine if new pest or disease management programs are needed. In addition, Congress directs the Food and Drug Administration, the Food Safety and Inspection Service, and state-level Departments of Agriculture to participate in inspecting many products regulated by APHIS.

APHIS and CBP personnel inspect shipments of imported agricultural products and certify that the required import health permits and SPS certificates accompany each shipment. One of the major flagship programs that APHIS and CBP administer together is the Agricultural Quarantine Inspection (AQI) program, in which APHIS and CBP technical staff work to ensure that the required animal or plant health permits, sanitary certificates (for animal products), and phytosanitary certificates (for plant products) accompany each shipment. APHIS transfers funds to CBP to conduct AQI activities.

The ongoing congressional commitment to preventing plant and animal disease and pests from entering the United States through agricultural imports is evident in annual appropriations Congress provides for APHIS. Congress has directed APHIS to monitor pests and diseases and has assigned APHIS to oversee SPS activities in some free trade agreements. Moreover, legislation introduced in the 115th Congress sought to address invasive species (e.g., Areawide Integrated Pest Management, H.R. 5411) and would have directed CBP to enforce APHIS regulations to deter smuggling of plants and animals into the United States.

INTRODUCTION

In the past two decades, increasing global trade liberalization, among other factors, has led to a rise in U.S. agricultural imports. The U.S. Department of Agriculture (USDA) reports that U.S. agricultural imports reached nearly $121 billion in the 2017 calendar year (*Figure 1*), which was twice the level of such imports in 2000.[1] Consequently, the increasing volume of imports has heightened public concern about the potential for introducing pests and diseases. Most imported agricultural products, such as live animal and plants, are inspected by U.S. government officials and have accompanying documentation—animal and plant health import permits—certifying adherence to U.S. requirements. More than ever, the U.S. government depends on the proper use of these import permits to facilitate U.S. agricultural trade.

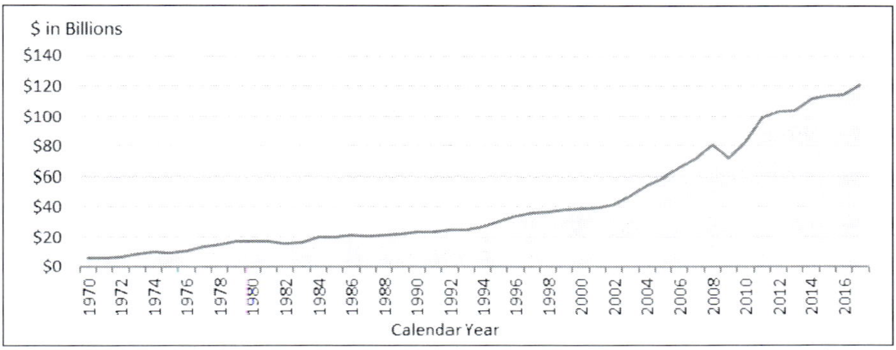

Source: USDA, Global Agricultural Trade System, and U.S. Census Bureau trade data. Accessed July 9, 2018.

Notes: Calendar year data. Values have not been adjusted for inflation.

Figure 1. Total U.S. Agricultural Imports, 1970-2017.

The USDA's Animal and Plant Health Inspection Service (APHIS) regulates the import, transit, and release of regulated animals, animal

---

[1] USDA, Global Agricultural Trade System (GATS) and U.S. Census Bureau trade data. Accessed July 9, 2018.

products, veterinary biologics, plants, plant products,[2] pests, organisms, soil, and genetically engineered organisms. APHIS-issued import health permits verify that the health status and the production practices of an imported product meet U.S. import standards. APHIS works with foreign exporters, U.S. importers, and foreign governments to interpret and enforce APHIS-issued animal and plant health import permits.[3]

Animal and plant health import permits include components from U.S. specific regulations and World Trade Organization (WTO) guidelines. Further, these health import permits are a part of broader agreements between the United States and its trading partners in the WTO that establish sanitary and phytosanitary (SPS) standards.[4] SPS measures aim to protect against diseases, pests, toxins, and other contaminants. Since 2000, the United States has entered into a dozen free trade agreements (FTAs),[5] which include an SPS chapter containing specific U.S. import requirements that the partner country has agreed to recognize. Examples include specific product or processing standards, requirements for products to be produced in disease-free areas, quarantine and inspection procedures, sampling and testing requirements, residue limits for pesticides and drugs in foods, and prohibitions on certain food additives.

APHIS works with the Department of Homeland Security's (DHS) Customs and Border Protection (CBP), in addition to other federal agencies (e.g., Food Safety and Inspection Service, Food and Drug Administration), to enforce agricultural import regulations. CBP has authority to enforce APHIS regulations at ports of entry.[6] APHIS and CBP personnel inspect shipments of imported agricultural products and certify

---

[2] Plant and plant product permits include plants for planting such as nursery stock, small lots of seed, and postentry; plant products such as fruits and vegetables, timber, cotton, and cut flowers; protected plants and plant products such as orchids and threatened and endangered plant species; transit permits to ship regulated articles into, through, and out of the U.S.; and controlled import permits to import prohibited plant materials for research.

[3] For more information, see CRS Report R45267, Animal and Plant Export Health Certificates in U.S. Agricultural Trade.

[4] For more information, see CRS Report R43450, *Sanitary and Phytosanitary (SPS) and Related Non-Tariff Barriers to Agricultural Trade.*

[5] For more information, see CRS Report R44044, *U.S. Trade with Free Trade Agreement (FTA) Partners.*

[6] DHS Agreement Number: BTS-03-0001, USDA-APHIS Agreement Number: 03-1001-0382-MU, "Memorandum of Agreement Between the DHS and USDA," February 28, 2003.

that the required animal or plant health import permits and SPS documentation accompany each shipment.

APHIS AUTHORITY OVER HEALTH IMPORT PERMITS

For much of the 20th century, animal and plant health bureaus within USDA operated independently of one another. The creation of APHIS consolidated these bureaus in 1972. There are a number of statutes (e.g., *Table 1*) that have established APHIS's authority over health import permits. However, the majority of the directives are found in two key legislative actions:

- The Animal Health Protection Act (AHPA, 7 U.S.C. §§8301 *et seq.*) is the primary federal law governing the protection of animal health. It gives APHIS broad authority to detect, control, or eradicate pests or diseases of livestock or poultry. AHPA consolidated all of the animal quarantine and related laws—some dating back to the late 1800s—and replaced them with one statutory framework. While most of the authorities contained in the consolidated AHPA were taken from existing laws, some new provisions were added to fill in gaps in legal authority.
- The Plant Protection Act of 2000 (PPA, 7 U.S.C. §§7701 *et seq.*) is the primary federal law governing plant pests in foreign and interstate commerce, covering agricultural commodities, plants, biological control organisms, articles that might be infested, means of transportation, and other pathways for moving pests. It authorizes APHIS to prohibit or restrict the importation, exportation, and interstate movement of plants, plant products, certain biological control organisms, noxious weeds, and plant pests.[7] It also authorizes APHIS to inspect foreign plant imports,

[7] Noxious weeds are undesirable plants that infest land or water resources and cause physical and economic damage. Under the PPA—which consolidated APHIS authorities, including almost all of the provisions of the Noxious Weeds Act (7 U.S.C. 2801 *et seq.*)—APHIS

quarantine any state or premise infested with a new pest or noxious weed, and cooperate with states in certain control and eradication actions.

Both AHPA and PPA give APHIS authority to inspect agricultural imports. However, after the events of September 11, 2001, congressional concern about agroterrorism—the deliberate introduction of an animal or plant disease to infect food, causing economic losses and/or undermining social stability—triggered the strengthening of APHIS and other federal agency agricultural inspection activities. Congress passed the Public Health Security and the Bioterrorism Preparedness and Response Act of 2002 (16 U.S.C. §§3371-3378; commonly referred to as the Bioterrorism Act) to bolster protection of the nation's food and water supplies and prevent unauthorized access to certain animal and plant disease organisms in laboratories. Since the enactment of the Bioterrorism Act, APHIS-issued health permits have been required to accompany APHIS-regulated agricultural imports to facilitate trade.

Annual Appropriations to Conduct APHIS Health Import Permit Activities

Overall, the House and Senate Agriculture Committees maintain jurisdiction over USDA's meat and poultry inspection programs and also other food-safety-related programs administered by other USDA agencies.[8] These House and Senate committees direct APHIS, other federal agencies, states, industry, and professional groups to facilitate international and domestic agricultural trade.

---

works to prevent noxious weeds from entering the country and conducts cooperative control/eradication programs with the states. Local governments frequently impose taxes on landowners to carry out noxious weed control programs.

[8] However, identifying committees of jurisdiction for specific laws, programs, and federal agencies is not straightforward and further complicated by split jurisdiction between FDA and USDA in the case of some foods due to documented fragmentation, overlap, and duplication among the agencies responsible for administering the laws and programs governing certain foods. For more background information, see CRS Report 98-175, *House Committee Jurisdiction and Referral: Rules and Practice*; and CRS Report 98-242, *Committee Jurisdiction and Referral in the Senate*.

Table 1. Select Statutes Concerning APHIS Authority Over Import Permit Issuance

General	
21 U.S.C. §§136-136a	Collection of fees for inspection services
7 U.S.C. §5925	High-priority research and extension initiatives
Animal Health	
7 U.S.C. §§8301-8317	Animal Health Protection Act
49 U.S.C. §80502	Transportation of animals (also referred to as the "28-Hour Law")
7 U.S.C. §2274	Firearm authority of employees engaged in animal quarantine enforcement
21 U.S.C. §§151-159	Virus-Serum-Toxin Act
Plant Health	
7 U.S.C. §§7701-7772; 7781-7786	Plant Protection Act of 2000
7 U.S.C. §§2279e, 2279f	Title V of the Agricultural Risk Protection Act of 2000
16U.S.C. 3371-3378	Lacey Act
7 U.S.C. §8401	Title II, Subtitle B, of the Public Health Security and Bioterrorism
7 U.S.C. §§281-286	Preparedness and Response Act of 2002 Honeybee Act
Animal Welfare	
7 U.S.C. §§2131-2159	Animal Welfare Act

Source: Compiled by CRS. FY2019 President's budget, APHIS, p. 2, https://www.obpa.usda.gov/20aphis2019notes.pdf.

The Subcommittees on Agriculture, Rural Development, Food and Drug Administration, and Related Agencies of the House and Senate Appropriations Committees appropriate funds for APHIS. Between FY2014 and FY2018, APHIS's discretionary appropriation has averaged $877 million annually. APHIS's appropriations cover four broad mission areas: (1) Safeguarding and Emergency Preparedness/Response, (2) Agency-Wide Programs, (3) Safe Trade and International Technical Assistance, and (4) Animal Welfare.[9] The Safeguarding and Emergency Preparedness/Response portion of the APHIS budget is responsible for monitoring animal and plant health in the United States and throughout the world and represents roughly 85% of APHIS's annual budget. Most of the animal and plant health import permit activities are housed in this mission area. In May 2018, both the House and Senate appropriations committees

[9] See CRS In Focus IF10953, *Agriculture Appropriations: Animal and Plant Health*.

reported bills (H.R. 5961, S. 2976, 115th Congress) that would have provided roughly $1 billion for the APHIS budget for FY2019. This amount was roughly $260 million more than Administration's FY2019 request and would have amounted to an increase of 7% from the FY2018 appropriation of $981.9 million.[10] In August 2018, the Senate passed H.R. 6147, which would have provided $1 billion for APHIS for FY2019.

APHIS-Issued Federal Orders

When the APHIS administrator considers it necessary to take emergency action to protect U.S. agriculture or prevent pest or disease entry into the United States, the administrator may issue a Federal Order. Such Federal Orders are effective immediately, contain specific regulatory requirements, and remain in effect until they are revised by another Federal Order or until an interim rule on the subject is published.

APHIS issues Federal Orders, under the PPA, which authorizes USDA to prohibit or restrict the importation or entry of any plant, plant part, or article that USDA identifies as necessary to prevent the introduction or dissemination of a plant pest into or within the United States.[11] For example, in June 2018, APHIS released a Federal Order prohibiting the importation of pomegranate arils[12] from Peru into the United States due to concerns about the potential importation of the Mediterranean fruit fly (*Ceratitis capitata*).[13] This Federal Order is still in effect as of January 9, 2019.

Obtaining an Animal or Plant Health Import Permit

Agricultural imports arrive to the United States through five U.S. Customs Districts (*Figure 2*). Although APHIS regulations are enforced at

[10] Plus $3.2 buildings and facilities costs.
[11] PPA, as amended, §412(a), 7 U.S.C. 7712(a).
[12] A pomegranate aril is a seed pod of a pomegranate.
[13] APHIS, Federal Import Order, June 11, 2018. The Mediterranean fruit fly can infest a wide range of commercial and garden fruits, nuts, and vegetables, such as apple, avocado, bell pepper, citrus, melon, peach, plum, and tomato.

the ports of entry, they are typically mediated through health import permits.

APHIS's Plant Production Quarantine and Veterinary Services oversee the health import permit process for a "plant health import permit" or "animal health import permit," respectively (*Figure 3*). APHIS works with several federal agencies to issue import permits.

U.S. importers obtain APHIS health import permits via the APHIS website ("ePermit") or in consultation with APHIS or state Department of Agriculture offices. Many U.S. importers use ePermit for the importation of products from abroad as well as for interstate trade. The ePermit system enables federal regulatory officials to issue, track, and verify the validity of import permits online. There is a cost to the importer associated with each ePermit application.[14]

APHIS and the state-level authorities often request health import permit submissions alongside supporting documentation (*Figure 4*).[15] In many cases, a biosafety facility inspection is a part of the review process—where a facility (e.g., laboratory, greenhouse, growth chamber) must demonstrate they can adequately and safely contain certain organisms.[16] Health import permits are normally processed and issued within 10 business days of receipt of the application.

After a health import permit is obtained, APHIS and/or CBP must inspect the APHIS-regulated product. One of the flagship programs that APHIS and CBP collaborate on and administer is the Agricultural Quarantine Inspection (AQI) program. AQI ensures that the required health permits, sanitary certificates (for animal products), and phytosanitary certificates (for plant products) accompany each shipment. APHIS transfers funds to CBP to conduct AQI activities.

---

[14] 9 CFR 130.2 and 7 CFR 354.3.

[15] APHIS, "ePermits," December 12, 2017, https://www.aphis.usda.gov/permits/downloads/ePermits_FAQs.pdf.

[16] Often conducted and reviewed by APHIS, the objective of the biosafety facility inspection is to verify that the facility has protocols in place to contain certain plant and animal pests or diseases (7 C.F.R. 330).

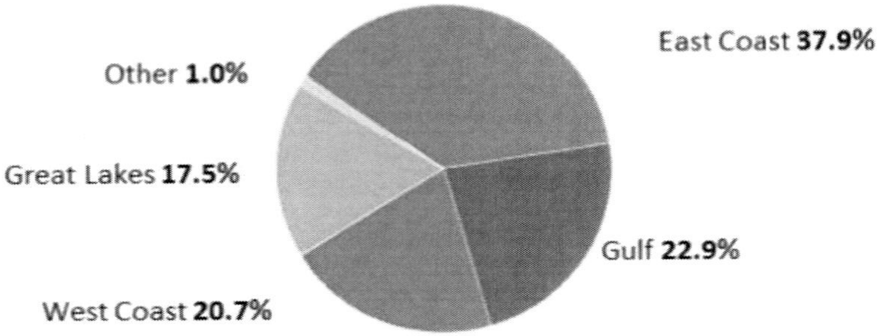

Source: USDA, GATS, U.S. Census Bureau Trade Data, U.S. Customs District Data, accessed July 2018.

Notes: 2017 calendar year data. East Coast Districts: New York, NY; Philadelphia, PA; Miami, FL; Savannah, GA; Norfolk, VA; Baltimore, MD; Tampa, FL; Charleston, SC; Boston, MA; Portland, ME; Charlotte, NC; Washington, DC; and Providence, RI. Gulf Districts: Laredo, T)(; Houston-Galveston, T)(; Nogales, AZ; New Orleans, LA; El Paso, T)(; St. Louis, MO; Mobile, AL; Dallas-Ft. Worth, T)(; and Port Arthur, T)(.West Coast Districts: Los Angeles, CA; San Francisco, CA; Seattle, WA; San Diego, CA; Great Falls, MT; and Portland, OR. Great Lakes Districts: Detroit, MI; Buffalo, NY; Pembina, ND; Chicago, IL; Ogdensburg, NY; Duluth, MN; Cleveland, OH; Minneapolis, MN; and Milwaukee, WI. Other Districts: San Juan, PR; Honolulu, HI; Virgin Islands; and Anchorage, AK.

Figure 2. Total U.S. Agricultural Imports by Customs District (2017).

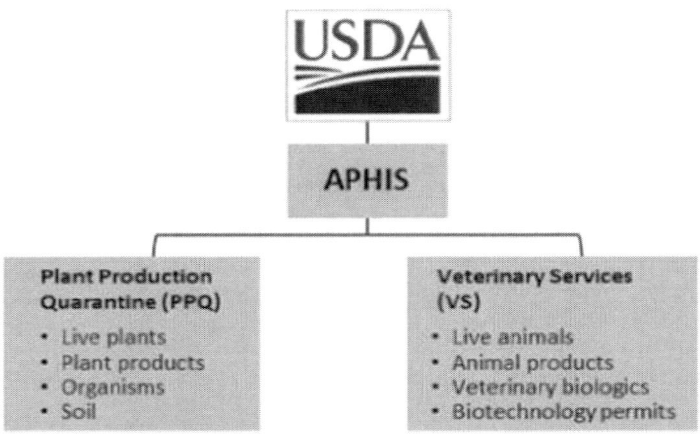

Source: CRS, using information from APHIS, Permits and Certifications, March 7, 2016.

Notes: APHIS cooperates with many federal agencies—including CBP, FDA, and FSIS—to facilitate the import permit issuing process. Veterinary biologics are vaccines, bacterins, diagnostics, etc., that are used to prevent, treat, or diagnose animal diseases. These products generally work through an immunological method or process.

Figure 3. APHIS's Health Import Permit Issuing Offices.

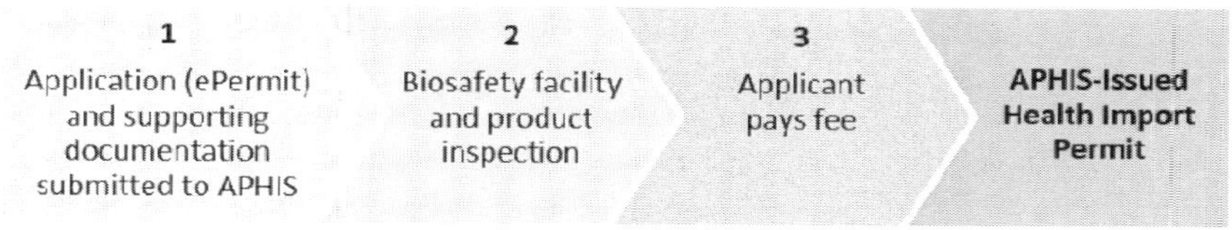

Figure 4. Flow Chart of APHIS-Issued Animal or Plant Health Import Permit.

APHIS collects AQI user fees from international airline passengers, operators of commercial vehicles,[17] cruise ship passengers, and importers of shipments requiring phytosanitary treatments.[18] Congress appropriates funding for AQI each year (e.g., operating expenses such as rent, utilities, travel, and supplies to conduct program activities). The AQI user fees recover costs that APHIS and CBP bear to administer the inspections. USDA estimates that in FY2018, AQI collected $765 million in fees, of which it transferred $539 million to DHS and retained $226 million to augment its discretionary appropriation.[19] In a 2013 report, the Government Accountability Office (GAO) identified a gap between fee revenues and total program costs.[20] In its report, GAO recommended aligning the division of fees between APHIS and CBP with their respective costs and that the two agencies ensure that fees are collected when due. In February 2018, APHIS announced it had implemented all of GAO's recommendations.[21]

ROLE OF OTHER GOVERNMENT ENTITIES IN PLANT AND ANIMAL IMPORTS

Importers must obtain APHIS-issued health import permits before beginning the inspection process. In addition to APHIS, Congress has

---

[17] Vehicles include aircraft, commercial maritime vessels, commercial trucks, and commercial railroad cars.

[18] Under Section 2509 of the Food, Agriculture, Conservation, and Trade Act of 1990.

[19] USDA FY2019 President's budget, APHIS, https://www.obpa.usda.gov/20aphis2019notes.pdf.

[20] GAO, "Agricultural Quarantine Inspection Fees: Major Changes Needed to Align Fee Revenues with Program Costs," GAO-13-268, March 2013.

[21] USDA FY2019 President's budget, APHIS, https://www.obpa.usda.gov/20aphis2019notes.pdf.

directed other federal agencies and state-level Departments of Agriculture (see Table 2) to participate in inspecting APHIS-regulated products:

- The Food and Drug Administration (FDA) has taken steps to protect the public from terrorist attacks on the U.S. food supply and other food-related emergencies in collaboration with DHS. Since 2003, FDA has advised U.S. importers to submit "prior notice" online forms[22] to FDA before food is imported or offered for import into the United States. Unlike APHIS, FDA does not enforce industry guidelines, but FDA does review APHIS-issued import permits to verify the disease or pest status of the agricultural product. The Food Safety Modernization Act (FSMA; P.L. 111-353) created new rules governing FDA's food inspection regime of both domestic and imported foods under the agency's jurisdiction.[23]
- CBP and APHIS inspect agricultural products together through the AQI program.[24] CBP officials require U.S. importers to present an APHIS-issued import permit before conducting inspections.
- The Food Safety and Inspection Service (FSIS) has regulatory oversight of meat, poultry, and some egg products. FSIS often requests APHIS-issued health import permits before proceeding with meat inspections. In the case of FSISregulated products, FSIS requires APHIS-issued animal health import permits to ensure that the meat and/or poultry ingredients in such food products are prepared under FSIS inspection or in a foreign establishment certified by a foreign inspection system approved by FSIS.[25]

---

[22] FDA, "Prior Notice Food Articles Quick Start Guide," July 16, 2018, https://www.fda.gov/Food/GuidanceRegulation/ImportsExports/Importing/ucm121048.htm.

[23] FSMA amends the Federal Food, Drug, and Cosmetic Act. For more information, see CRS Report R43724, *Implementation of the FDA Food Safety Modernization Act (FSMA, P.L. 111-353)*. FSMA was the largest expansion of FDA's food safety authorities since the 1930s.

[24] See "Introduction" for a discussion on CBP activities.

[25] APHIS, "Changes to USDA APHIS VS Import Permitting Process for USDA, FSIS Exempted Food Products Containing Meat and Poultry Ingredients," March 31, 2009.

- State-level Departments of Agriculture enforce APHIS regulations. States have different animal and plant health import restrictions that apply to interstate trade. Some states have heightened restrictions based on disease and pest detections. Similar to FSIS, state-level regulators typically request APHIS-issued animal or plant health import permits before conducting their inspections. State-level Departments of Agriculture often coordinate with APHIS and CBP directly (e.g., when a state detects a disease outbreak).

Table 2. Select Government Entities that Inspect Agricultural Imports

Government Entity	Commodity Jurisdiction	APHIS Health Import Permit Enforcement
APHIS	Animals, animal products, veterinary biologics, plants, plant products, pests, organisms, soil, and genetically engineered organisms	Yes
Food and Drug Administration (FDA)	Dietary supplements, bottled water, food additives, infant formula, other food products (not regulated by USDA), eggs in the shell, and game meats	No
Customs and Border Protection	All APHIS-regulated commodities	Yes
Food Safety and Inspection Service	Meat, poultry, and some eggs	Yes
State-Level Departments of Agriculture	All APHIS-regulated commodities	Yes

Source: Summarized by CRS. For more information on the agencies and their roles see CRS Report RS22600, The Federal Food Safety System: A Primer.

Notes: For a full listing of FDA regulated products see FDA, Regulated Products, August 3, 2018, https://www.fda.gov/ForIndustry/ImportProgram/ImportBasics/ RegulatedProducts/default.htm.

CONGRESSIONAL ACTIVITIES

Congress has long been involved in efforts to prevent the entry of pest and disease threats into the United States from agricultural imports. This oversight has manifest in various congressional actions, including continuing to provide APHIS with appropriations to monitor pests and diseases, introducing legislation to address invasive species (e.g., Areawide Integrated Pest Management, H.R. 5411, 115th Congress), and directing CBP to enforce APHIS regulations to deter agricultural smuggling into the United States. APHIS-issued animal and plant health import permits can be a tool to prevent agricultural import threats to the United States. The sections that follow summarize selected issues that Congress has addressed by directing APHIS to undertake a specific role, assigning the agency with specific responsibilities, or authorizing it carry out certain actions.

Pests and Diseases

In the event of a U.S. disease outbreak or pest infestation, APHIS is designated to be the lead U.S. agency for informing the international community (such as to the World Organization for Animal Health, also known as "OIE").[26] Typically, APHIS would directly contact the partner country's scientific authority to explain the nature and extent of the outbreak. In most cases, APHIS and officials of the partner country observe the SPS guidelines agreed upon under the WTO framework or, in some cases, the SPS chapters in individual FTAs.[27] Responses to a pest or disease outbreak can include a complete ban on the importation of U.S. products impacted by the pest/disease. Under certain circumstances, a "regionalization" protocol may be applied in which a specific exporting region within the United States may be recognized as disease- or pest-free.

In the event of a trading partner experiencing a disease outbreak or pest infestation, the notification process is similar to a U.S. outbreak (e.g.,

---

[26] OIE is based in Paris and known by its former French abbreviation, OIE (Office International des Épizooties), even though it has been renamed the World Organization for Animal Health. OIE is not part of the United Nations but is a separate intergovernmental organization that was founded in 1924 under an international agreement. For more information, see OIE's website: http://www.oie.int/index.php.

[27] APHIS is the lead federal agency charged with enforcing SPS chapter agreements in FTAs as well as WTO SPS guidelines/commitments in order to limit the disruption to agricultural trade from pest and disease detections or outbreaks. See WTO, "Sanitary and Phytosanitary Measures," https://www.wto.org/english/tratop_e/sps_e/sps_e.htm.

informing OIE). APHIS informs U.S. importers of the trading partner outbreak. U.S. importers work with APHIS to verify that the health import permit reflects the disease or pest status of the exporting country. APHIS provides guidance to U.S. importers if their health import permits would be accepted (e.g., following SPS protocol of regionalization) or rejected (e.g., ban products from an infected trading partner).

Table 3. Select APHIS-Monitored Pests and Diseases

Plant	
• Asian citrus psyllid	• Imported fire ant
• Asian gypsy moth	• Khapra beetle
• Asian longhorned beetle	• Light brown apple moth
• *Citrus greening*	• Mediterranean fruit fly
• Coconut rhinoceros beetle	• Mexican fruit fly
• Emerald ash borer beetle	• Old world bollworm
• European cherry fruit fly	• Oriental fruit fly
• False codling moth	• Spotted lanternfly
• Giant African snail	• *Sudden oak death*
Animal	
• *Avian influenza*	• New World screwworm
• *Bovine spongiform encephalopathy*	• *Piroplasmosis*
• *Brucellosis*	• *Porcine epidemic diarrhea*
• Cattle fever tick	• *Pseudorabies*
• *Chronic wasting disease*	• *Schmallenberg virus*
• *Classical swine fever*	• *Scrapie*
• *Contagious equine metritis*	• *Spring viremia carp*
• *Equine herpesvirus*	• *Swine influenza*
• *Equine infectious anemia*	• Ticks
• *Equine piroplasmosis*	• *Tuberculosis/Brucellosis*
• *Equine viral arteritis*	• *Tuberculosis*
• *Foot-and-mouth disease*	• *Vesicular stomatitis*
• *Johnes*	• *West Nile virus*

Source: APHIS, "Plant Pests and Diseases Programs" and "Animal Disease Information." January 30, 2018.

Notes: Italicized font refer to diseases. Nonitalicized font refer to pests.

Table 3 provides a selected listing of prominent APHIS-monitored diseases and pests that pose a threat to U.S. agriculture. Among these are disease concerns, including avian influenza and foot-and-mouth disease, and invasive plant pests such as the Asian longhorned beetle and the emerald ash borer. In some instances, agricultural imports (or even interstate shipments) can arrive into AQI with pests and diseases that could impact public health or U.S. agricultural production systems.

Congress has sometimes directed APHIS to address certain diseases through the annual appropriations process. For example, the Consolidated Appropriations Act, 2018 (P.L. 115-141), includes a general provision providing APHIS with an additional $5.5 million—to remain available until the end of FY2019—to fund a multiple-agency response to citrus greening disease.[28]

Agroterrorism

Agroterrorism is the deliberate introduction of an animal or plant disease with the objective of infecting food, causing economic losses and/or undermining social stability. Requiring U.S. importers to obtain APHIS-issued health import permits is one way that APHIS and the CBP protect against agroterrorism. Permits provide APHIS and CBP inspectors an opportunity to conduct random sampling to assist in disease and pest identification or to detect other potential threats.

Congress has taken steps to address potential harmful imports ("select agents") that could impact public health or animal/plant health[29] through the Bioterrorism Act by directing the Department of Health and Human Services's (HHS) Centers for Disease Control and Prevention (CDC) and APHIS to enforce the *Select Agents and Toxins List* under the Federal

---

[28] P.L. 115-141, §771. Citrus greening, also called Huanglongbing or yellow dragon disease, is spread by an insect called the Asian citrus psyllid. The psyllid feeds on the stems and leaves of the trees, infecting the trees with the bacteria that causes citrus greening. Greening impairs the tree's ability to take in nourishment, ultimately resulting in fewer and smaller fruit over time. There is currently no cure for infected trees.

[29] For more information, see section "Microbial Pathogens in the Laboratory: Safety and Security" in CRS Report R44786, *Science and Technology Issues in the 115th Congress*.

Select Agent Program.[30] The agents and toxins on this list have been determined to pose a threat to human and animal health, plant health, or animal and plant products.[31] Some of these agents and toxins are overseen by HHS, others by USDA, and certain ones by both agencies.

Invasive Species

An invasive species is a nonnative (also known as an *alien*) species that does or is likely to cause economic or environmental harm or harm to human health.[32] Invasive species include plants, animals, and microbes. The introduction of invasive species into the United States—whether deliberate or unintentional—can threaten native animal and plant communities, lead to ecosyst6em disruptions, and contribute to extinctions of native species. Invasive species can also impact biodiversity and alter habitats and can result in the introduction of new pests and diseases.[33] An estimated 50,000 non-native invasive animal and plant species have been introduced to the United States for over a century, with a 2011 study citing total costs exceeding $100 billion annually— including economic costs related to damages as well as management, mitigation, and recovery activities.[34]

APHIS collects information on invasive species and monitors their impacts (e.g., on agricultural production, forest lands). The information is updated and shared with AQI and ports of entries. The health import permits provide the APHIS/CBP officials specific guidelines (e.g., identifying specific pests), depending on the origination of the product, that facilitate detection during inspecting or random sampling of the imported

---

[30] HHS and USDA Select Agents and Toxins, 7 CFR Part 331, 9 CFR Part 121, and 42 CFR Part 73.

[31] An attenuated strain of a select agent or an inactive form of a select toxin may be excluded from the requirements of the Select Agent Regulations.

[32] For more information, see CRS In Focus IF11011, *Invasive Species: A Brief Overview*.

[33] See CRS Report R43258, *Invasive Species: Major Laws and the Role of Selected Federal Agencies*.

[34] Inherent in any calculation of the costs of invasive species, however, is valuation of economic and societal factors on which expert opinion differs. See D. Pimentel, "Environmental and Economic Costs Associated with Alien Invasive Species in the United States," in *Biological Invasions: Economic and Environmental Costs of Alien Plant, Animal, and Microbe Species* (Abingdon, UK: CRC Press, 2011).

agricultural products. U.S. importers also work with APHIS and other federal agencies (e.g., the Environmental Protection Agency) to obtain regulations on preventing an invasive species infestation (e.g., vessel ballast water discharge).[35]

Several statutes provide federal agencies with authorities to address invasive species in the United States. Some in Congress have expressed interest in pursuing legislation to reduce the prevalence of invasive species, such as through the introduction of "area-wide integrated pest management" (e.g., H.R. 5411, 115th Congress), a pest management strategy that is applied within a geographical area. This bill would have expanded the USDA Integrated Research, Education, and Extension Competitive Grants Program for qualified area-wide integrated pest management projects.

Smuggling

Over the past 30 years, there has been a steady increase in the movement of people and agricultural products around the world.[36] The volume of smuggled and improperly imported agricultural products entering the United States has been a congressional concern.[37] Products smuggled into the United States can harbor exotic plant and animal pests, diseases, or invasive species that could damage domestic crops, livestock, and the environment. If APHIS identifies an illegally imported product or a regulatory violation, it may seize the item and pursue civil and criminal penalties, if warranted. APHIS encourages distributors and retailers to purchase products that have been imported through legal channels that are typically accompanied by APHIS health import permits.

One way that Congress has attempted to prevent agricultural smuggling is through the Lacey Act.[38] The Lacey Act dates from 1900 and prohibits the importation of any plant—with limited exceptions—that is

---

[35] See CRS Report R42142, EPA's Vessel General Permits: Background and Issues.

[36] For more information see D. Reynolds et al., "Smuggling of Live Animal and Food: Current Practice, Prevention and Repression Tools," World Organization for Animal Health, 2006, pp. 137-142.

[37] USDA FY2019 President's budget, APHIS, https://www.obpa.usda.gov/20aphis2019notes.pdf.

[38] 16 U.S.C. §§3371-3378.

taken or traded in violation of domestic or international laws.[39] The act requires declarations—in addition to the APHIS-issued plant health import permits—for imported shipments of most plants or plant products. APHIS works with CBP, the U.S. Coast Guard (e.g., for fisheries violations), the National Marine Fisheries Service, the Federal Bureau of Investigation, the U.S. Forest Service, and U.S. Immigration and Customs Enforcement to enforce the Lacey Act through inspection or monitoring activities. APHIS's role in this context is to collect APHIS-issued health import permits, manage the declaration requirement, provide guidance to importers regarding the declaration, perform compliance checks, and provide enforcement agencies with information to assist their investigations. The declaration is to be made by the importer at the time of import. According to USDA, both CBP and APHIS activities have contributed to an increase in the number of declarations, with approximately 1 million declarations in FY2017, up roughly 300,000 from FY2016.[40] This increase has coincided with the introduction of the online system in lieu of paper-based declarations.

---

[39] For more information, see CRS Report R42119, The Lacey Act: Compliance Issues Related to Importing Plants and Plant Products.

[40] USDA FY2019 President's budget, APHIS, https://www.obpa.usda.gov/20aphis2019notes.pdf.

INDEX

A

B

C

D

E

F

G

H

I

J

L

M

N

O

P

T

Related Nova Publications

New Developments in Agricultural Research

Editor: Patrik Blanchard

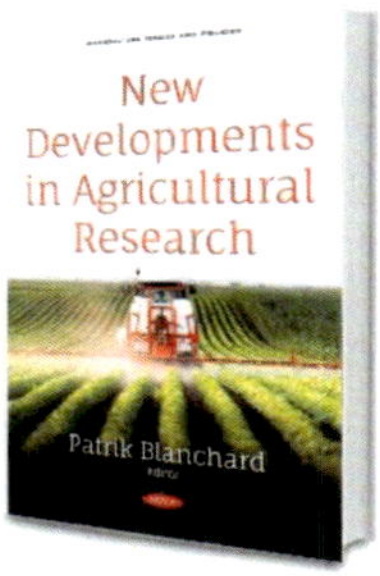

Series: Agriculture Issues and Policies

Book Description: *New Developments in Agricultural Research* provides a comprehensive introduction and overview of portable MMSs applied to agricultural and forestry, to highlight the potentialities and challenges of this novel technology in this specific application field.

Softcover ISBN: 978-1-53615-363-7
Retail Price: $95

Agricultural Policy, Appropriations and Discretionary Spending

Editor: Albertus Brinkhuis

Series: Agriculture Issues and Policies

Book Description: This book provides information on the 2018 and 2019 discretionary spending as well as a copy of the 2019 Agriculture appropriations act.

Hardcover ISBN: 978-1-53615-215-9
Retail Price: $160